THE KING OF TRICKS

MANIPULATION OF CLINICAL DRUG TRIALS

A Novel by John Gill

AF719608

Copyright © 2025 A Novel by John Gill
All rights reserved.

No part of this publication may be reproduced in any form or by any means, electronic or mechanical, including photocopying, recording, or any information browsing, storage, or retrieval system, without permission in writing from John Gill.

For Monty Lyle Gill

About the Author

John Gill is a retired federal biologist who led teams of biologists, engineers, hydrologists, archeologists, in preparing environmental impact statements on federal projects such as hydroelectric plants, nuclear power plants, and ballistic missile bases.

Other books by John are:

- *The Mind Doctor: A Revealing Portrayal of Psychopharmacological Corruption*
- *The Wannabe Doctor: A Man's Quest Amid Psychiatry's Identity Crisis*

Historically, medicine is the only science that conducts life-threatening experiments on human subjects in order to advance its knowledge base in the name of progress. When progress and commercialization coincide, the potential for abuse is great.

— Theresa Richardson, 2003
Book Review of *Human Experiments at the Holmesburg Prison*, 1998

Contents

Chapter 1: The Issue

> The person who takes medicine must recover twice, once from the disease and once from the medicine.
>
> — William Osler
> Canadian physician
> 1849-1919

In July 1990, crowds lined both sides of Center Street in Salem, Oregon. A homemade sign said, "Fremontria is poison."

TV newscasters interviewed people on both sides of a wrongful death case. People around the world were following *Sanders versus Hancox Pharmaceuticals.* Four people had died in Tangent, Oregon. Public sentiment was divided over the risks of psychotropic drugs.

An activist said, "For the drug maker, it's profits over patients."

Across the street, a man said, "Where would we be without drugs like Fremontria? We can't go back to life without them. We need modern medicine."

The federal courthouse was the center of attention. The largest courtroom was filled. A sign on the door said, "Court in session."

Closing arguments began. The plaintiffs' attorney, Daniel Hammond, spoke first.

"Hancox Pharmaceuticals doesn't make medications to benefit humanity. The company wants its drugs to work, but if Fremontria makes money, Hancox wins, regardless of whether patients live or die. The company is not our benefactor. It's in the drug business for the money.

"Through the court's discovery process, the plaintiffs acquired Hancox's internal documents. We showed them to you during the trial. Memos revealed that the pharmaceutical giant knew eight years ago that, in some cases, Fremontria is lethal. We would never have seen these documents if Tom Alexander and Sara Morgan had not brought this case. It's like what happened to Big Tobacco. The key was getting the documents that the tobacco companies had been hiding. The world got a new perspective when internal documents were made public.

"Hancox placed Fremontria on the market in 1989. We don't know exactly when the company first realized that the drug could induce homicide. Still, a memo from the company's chief of research to its chief of marketing says company officials were aware of the danger by 1987. Hancox hid that fact."

"Objection," a Hancox attorney said.

"Denied," the judge said, "This is Attorney Hammond's closing statement. I don't allow objections to closing arguments. Resume your presentation, Attorney Hammond."

"I lost a suicide case to Hancox a year ago. The company tricked me into arguing on an either-or basis. Was the cause of suicide the depression or the drug? They tricked me into fighting their kind of fight. They used rehearsed arguments while I broke new ground. Don't let Hancox succeed with the same drug-versus-disease argument in today's case. Frame the debate differently. Ask yourself if it's possible for the compound to produce a violent reaction in some people. Ask yourself why the drug's packaging doesn't warn about its ability to cause homicide and suicide.

"Under cross-examination, Hancox's chief of marketing admitted the company doesn't know the exact cause of depression. Without that knowledge, they cannot know how to treat the disorder chemically.

"Societal problems contribute to depression. Problems include poverty, homelessness, unemployment, bullying, divorce, and the death of a loved one. Doctors don't know if there is an exact medical cause, such as a biochemical condition.

"Fremontria is a 'me-too drug.' To obtain the FDA approval for sale in the US, Hancox used the Nelson Company's Zonall drug as a model and modified the chemical formula slightly. The FDA has lenient standards for approving me-too drugs, making them lucrative for pharmaceutical companies to manufacture. Drug companies invest in 'me too drugs' and rarely do innovative research. So, Hancox got in on the action by creating its own anti-depressive drug.

"Don't fall for the lie that Fremontria couldn't be at fault because the research was so great. The company didn't find a new molecule. It simply took some of the market share from the Nelson Company by modifying Zonall's chemical formula slightly. The research was mundane."

After a recess, Hancox's lead attorney, Roberto Alfonzo, made his closing argument.

"This case should never have come to trial. Raymond Sanders shot and killed his wife, his daughter, his grandson, and himself. That was tragic. The ridiculous part of the story is when the plaintiffs claim that Mr. Sanders did these things because he took two Fremontria pills in two days. Fremontria is a safe antidepressant used by millions of people. Two pills could not have caused three murders and a suicide.

"Mr. Sanders' depression caused this tragedy. Sadly, Fremontria didn't have time for its therapy to work. With time, it could have saved four lives.

"Mr. Sanders suffered three documented bouts of depression over the years. He ignored the advice of psychiatrists by not retiring from his stressful job as an anesthesiologist. He became agitated over everyday matters. For example, he threatened to kill a neighbor over a property line issue. Depression and Mr. Sanders' impaired judgment caused him to kill his family members and himself."

***

The jury selected its foreman. Sam Jenkins was a retired college instructor who had taught research methods.

The jury began to deliberate. Could Hancox be as cunning as the plaintiffs' witnesses claimed? It strained the imagination to think that two pills could have caused Raymond Sanders to kill three family members and himself. The jury argued among themselves with little regard for the evidence.

The juror who held the floor talked on and on.

A male juror said, "Let someone else talk."

Sam listened. Voices grew loud.

Irene Walters, a hairdresser, said, “Sam, you’re the foreman, and you taught about the kind of issues we’re haggling over. But you haven’t said where you are regarding Hancox’s culpability.”

Sam said, “For most of us, this trial is the first time we’ve dealt with a major medicolegal issue. We’re under an international spotlight. We need to get this right. We need to take our time, listen, and think —really think. We must cast aside our biases. We must respect our fellow jurors’ opinions. I don’t want to be your foreman unless we do these things.”

Sam stopped talking. He sat there. The other jurors looked at him. He sat there.

The talkative juror said, “Silent Sam won’t say a word. I volunteer to be the foreman. I’ll speak up.”

Clarence Curtis, a city fireman, said, “Sam is making a point. I’d rather have a thoughtful leader than someone like you who just runs his mouth. Sam, do something.”

“I am doing something. I’m showing you that I won’t go along with the loudest juror. I want you guys to respect each other, to think, and to honor the evidence.”

Clarence said, “I see what you’re doing. Let’s think instead of reacting emotionally. Right, Gang?”

A cheer went up. A bailiff opened the door and looked in. She smiled and closed the door.

Sam said, “All right, you’ve decided, and the decision applies to me too. Correct me if I disregard any evidence.”

The jury reviewed the evidence. They considered whether Hancox would sell a dangerous drug. The company's internal memos were incriminating.

Clarence said, "We can't ignore the evidence, even though Hancox is a big shot. The company seems to be saying, 'Don't believe what is evident, what is obvious. Don't believe a concept unless it's been found statistically significant.'

"But Fremontria was found in Mr. Sanders' blood. And he did things unlike anything he'd ever done. Hancox has an immense ability to influence doctors and patients. The company is a master of publicity, which explains how they've made billions in drug sales. It uses Madison Avenue techniques to market its drugs, and it's using those same techniques to persuade us."

Irene said, "TV ads say to ask your doctor. So I did. Doctors gave me whatever prescription I asked for every time. This happened with different doctors and different drugs."

Sam said, "Here's another factor that you should take into account: individual differences among people. Some adults are as short as four feet or less, while others are as tall as seven feet or more. Most people are in between.

"What does this have to do with how Mr. Sanders may have reacted to Fremontria? Some people may tolerate Fremontria, while others may have no tolerance for it. From my years as a biologist, I realize that individuals taking antidepressants will have varying responses. The health of some may improve slightly. Some may get a lot better. Some may get a little worse. And some may get a lot worse. Hancox should have warned healthcare officials of the perils of their drug. They should have written a more informative label for Fremontria."

After more back-and-forth, the jurors came together. Sam told a bailiff that the jury had reached a decision. The bailiff reported to the judge. She came back and said the jury was to return to the courtroom.

Attorneys, reporters, and spectators were already in the courtroom. The judge called the trial back into session. He looked at Sam.

"Has the jury reached a decision?"

"We have, Your Honor."

Sam handed a paper to a bailiff, who gave it to the judge. He looked at the verdict and returned it to Sam.

"Please read the verdict aloud."

Sam read, "The jury finds for the plaintiffs."

A gasp went through the courtroom.

The judge said, "Order in the court."

Sam resumed reading.

"The defendant is 95 percent responsible for the killings. Raymond Sanders is 5 percent responsible. The jury recommends that the defendant pay $12 million to be divided equally between the two plaintiffs."

The judge said, "I find the evidence convincing. I approve the jury's recommendation."

Handshaking and hugging surged throughout the courtroom.

***

Twenty minutes later, outside the courthouse, Attorney Alfonso spoke at a news conference.

"Facts, science, and common sense say that two Fremontria pills could not have caused this tragedy. It's a travesty that this case made it to trial, much less won a decision. Mr. Sanders' depression caused the four deaths. Hancox will appeal."

The plaintiffs' attorney, Daniel Hammond, spoke at a different press conference.

"The jury found that two pills caused four deaths. Why did the FDA approve Fremontria for the marketplace? That question applies to hundreds of drugs and is a vital medicolegal problem that must be investigated in depth."

Chapter 2: Boyhood Friends

> You will not be forgotten. A hundred years from now; men will be sitting around campfires and telling legends of the Lionheart.
>
> — Sharon Kay Penman
> American historical novelist
> 1945-2021

From when he was six, Richard Martinson helped his father, Ethan, who worked for a grass seed farmer in Oregon's Willamette Valley.

With time, Richard assumed more responsibility. He fertilized plants and harvested seeds. He loaded the seeds onto trucks for transport to local granaries or markets.

***

From age seven, Richard admired a legendary, fictional king of early England. He read *King Arthur and his Knights of the Round Table*, a retelling of many tales, so they became one story. He read John Steinbeck's *The Acts of King Arthur and His Noble Knights.*

Richard also read *A Connecticut Yankee in King Arthur's Court*. In Mark Twain's story, Hank Morgan, a nineteenth-century engineer in Connecticut, was hit on the head. Somehow, the blow transported him to England during the reign of King Arthur.

Hank was confused. Then he realized he was in sixth-century England. He used his knowledge to make people think he was a magician with great power. He ran afoul of Merlin, the established wizard. Hank tried to teach the people modern ways.

The Catholic Church tried to remove Hank from England. But he shot his enemies, using revolvers and gunpowder that he had developed.

Hank accomplished his feats because he had more knowledge than people of the sixth century. Hank's success caused Richard to vow that he would learn as much as he could and use that knowledge to make a fortune.

One day, in a grass seed field, Ethan noticed that Richard was daydreaming.

Ethan said, "King Arthur was a fictional character of sixth-century England. But Richard the First was a real king of the twelfth century. Folks called him Lionheart because he was so brave."

"Tell me about him."

"He led the Third Crusade. You should switch your attention from Arthur to Richard, who was a historical figure. My family descended from him. That's why I named you Richard."

Richard's chest swelled with pride. He was named after a king. He read about Lionheart. There was no doubt the king was brave.

But the boy was troubled. In 1191, Christian crusaders captured Acre, a city in Israel. Richard ordered the beheading of more than 2,500 Muslims. The massacre was done on a hill called Ayyadieh so Richard's enemies could see the slaughter. Most of the victims were male prisoners of war, but women and children also were killed.

Richard searched for facts that might justify Lionheart's cruelty. The king was at war. Saladin, the enemy sultan, had not kept his agreement for a prisoner swap. If the swap had taken place, the Lionheart would not have had so many prisoners. He had no place to keep them. He and his army would have to move on, and they had no way to take so many prisoners with them.

To resolve the turmoil in his mind, Richard researched and rationalized his thoughts. He wanted to have a hero. He created excuses for King Richard's cruel act, enabling the boy to idolize his namesake again.

Bryce Jewel lived a mile from Richard. The Jewels were shepherds. Bryce's best friends were border collies. Bryce and Richard went to school together. They met at the school bus stop, rode the bus, spent time in class, ate lunch, attended another class, and returned home.

***

It was a cold, drizzly Saturday morning in November 1947. Richard and Bryce were seven years old. They were outdoors, engaged in a make-believe war. What they thought they knew of history was a mix of facts, fiction, and imagination. This morning, they were part of the Third Crusade, attempting to capture Jerusalem from Saladin.

Ethan had helped the boys make two bows. The arrows were sticks with one end notched and the other sharpened. They had no feathers, but for the boys, they were good enough.

Sticks also were the horses that the boys rode into battle. One end of a stick dragged the ground as a boy straddled it and darted about. In their minds, the boys rode great muscular stallions.

Richard was the leader, for he was Lionheart. He determined what role Bryce would play. Sometimes Bryce was a soldier in Richard's army, and sometimes he was an enemy soldier.

Bryce said, "Why can't I be in charge sometimes?"

"Because I'm the king. Some people are born to lead and others to follow."

Richard's mother, Flora, called the soldiers in for soup.

Richard said, "I wish Dad were here to tell us more about King Richard."

"Maybe I can help," Flora said. "When King Richard was in England between battles, he met Robin Hood in Sherwood Forest. Robin became his friend."

"Great," Richard said.

"Was Robin Hood real?" Bryce said.

"There was probably no one person who was Robin Hood, but there's evidence that some of the stories happened. Maybe the things that different men did over time were lumped together if they were for the peasants and against the crooked government. Instead of referring to what five or six different men did, folks might have just rolled them into the deeds of Robin Hood. They may have needed a hero so much that they invented one."

"Did stories say Robin Hood met Lionheart?" Richard said.

"There is a story of Lionheart spending time in Sherwood Forest. It's said that he took part in archery contests and wrestled some of Robin's men. He feasted and drank with them before going back to his own army. You can put Robin in your stories of King Richard."

"I want to be Robin," Bryce said.

"That's okay," Richard said. "Lionheart is still the boss."

Chapter 3: An Accident that Changed a Life

Falls lead to nearly half of the traumatic brain injuries.

---Centers for Disease Control and Prevention
Traumatic Brain Injury & Concussions, 2022

It was raining. Richard and Bryce were playing in a granary. The seeds had been hauled to market, making room for adventures to unfold. To the boys, this was a castle where the Sheriff of Nottingham stored the riches he had taken from the peasants.

There were stairs and a loft. The loft was the sheriff's office. The little soldiers intended to return the tax money to the peasants. They fought their way to the top of the stairs.

The sheriff's men outnumbered Lionheart and Robin ten to one. But they were no match for the two from Sherwood Forest. Richard ordered the sheriff to turn over the money and valuables.

A Sheriff's man got behind Robin and swung his sword. Richard twirled to parry the thrust. He fell backward and skidded on his back down the stairs. His head clunked against each step. He came to rest with his feet above his head. Blood trickled from his nose.

Bryce ran to a field where Ethan and Hiram, Bryce's dad, were filling a fertilizer tank.

"Richard fell in the granary."

They got into Ethan's pickup, drove to the granary, and climbed the stairs. Richard lay still.

Ethan said to Hiram, "I'll hold his head so it's stable while you carry him down to the floor."

On reaching the floor, the men laid Richard down. Ethan and Hiram kneeled beside the boy. Hiram sent Bryce home.

Richard's nose was bleeding, and his eyes were closed. His eyelids began to twitch. He woke up.

"I'm okay."

Richard tried to stand but crumpled to his knees. He vomited.

The men carried Richard to the pickup. Hiram rode in the pickup's bed to give the boy room in the cab. It began to rain.

They arrived at Mercy Hospital in Lebanon. A technician x-rayed Richard's head and spine. Dr. Ralph Veres took charge of the case. He had served as an Army surgeon against Hitler's war machine. Veres' team included Nurse Gloria McGovern, who had also served in World War II.

Dr. Veres said, "We need to decompress the brain so it doesn't press on the brain stem."

Gloria shaved Richard's head. Veres drilled holes in the boy's skull. Using a saber saw, he connected the holes. He lifted out a piece of skull and handed it to Gloria.

"Store that in the freezer. If Richard is lucky, we'll put it back in place."

Richard spent two weeks in Mercy Hospital. He was solemn as he recovered. After being discharged from the hospital, he did not play with Bryce. When the doctor next saw him, he tested Richard's mental acuity.

Dr. Veres met with Ethan and Flora.

"Richard lost some memory, but I'm not worried as long as he can remember things that'll happen in the future. What worries me are changes in his behavior."

Flora said, "He doesn't play with Bryce."

"That's one of the things, but not the worst."

Ethan said, "He believes he's Richard the Lionheart. I heard him referring to himself as the king, but he shut up when he saw me."

Dr. Veres said, "I'm glad that Richard realized that you don't believe he's an ancient king. That indicates he has a concept of reality."

"Or it might mean he slips in and out of believing he's King Richard," Flora said. "Maybe sometimes he thinks he's a king, and sometimes he doesn't. Is there anything we can do to influence his beliefs?"

"Don't let Richard pressure you into treating him like he's Lionheart. Don't let him persuade you that he's royalty, even if he seems to be doing it in jest."

"How could he do that?" Ethan said.

"If you try to humor him, he'll take more privileges. You'd be reinforcing his belief. You've got to pull him back to reality."

Flora said, “It’s complicated because he’s a great pretender. He might think he’s King Richard, but that it’s best to deceive us into thinking he’s a twentieth-century boy.”

Chapter 4: Laura

> The question is not, "Can they reason?" nor "Can they talk?" but rather "Can they suffer?"
>
> — Jeremy Bentham
> English philosopher
> 1747-1832

Because they lived in a sparsely populated area, Richard and Bryce saw only a few people every week. During the school year, they saw their classmates and teachers. On the Sabbath, they saw their minister and their Sunday school teacher. During the summer, they worked with Ethan and Hiram in grass-seed farming. And they saw their mothers every day.

The only other person Richard saw regularly was his sister, who was four years his senior. She loved animals. Her favorites were two Giant Canada Geese. Maddie weighed 13 pounds, and Max weighed 17. The pair showed affection for Laura by honking, peeping, and nibbling at her shoes and pants cuffs.

The birds had hatched on the farm. Both had imprinted on Laura as she was the first thing they saw upon hatching. They followed her everywhere. Whenever Laura was outdoors, the three were inseparable.

The Martinsons held a get-together and invited family and friends. They prepared a feast by killing and cooking some of the farm animals.

The main dish turned out to be Max. Ethan caught the gander by hooking one of his feet. He didn't notice that Laura was watching when he cut off Max's head. She ran into the woods, fell to the ground, and wept.

The next day, Flora said, "Laura, you didn't eat any of the roasts. We can't afford to feed expensive meat like that to the dog."

"I wasn't hungry."

"Is it my cooking?"

"You're a good cook."

"Then what is it?"

"I see things differently than you and Dad."

"What's bothering you?"

"I want Dad to stop killing our animals and you to stop cooking them. Max is my brother as much as Richard. The only difference is that Dad killed Max."

"We eat to live, and we get protein from meat."

"I love all the animals we have. They're like people. Each has a personality. I've named every goose, goat, and chicken. Why do we kill them? Can we give up meat and eat only vegetables? That way, no one dies."

Flora walked to the table, sat down, and stared at Laura.

"Mom, you're trembling."

"We're farmers, Honey. We eat meat."

"We could sell eggs, milk, and vegetables. We don't have to kill and eat animals."

"We're farmers. You need to understand that. Where did you get this idea of not eating meat? Did someone at church get to you?"

"I decided on my own."

"Eating meat is not sinful. God gave us animals to use as we see fit. What you're suggesting isn't practical."

"The Christian Bible doesn't prohibit eating animals, but it doesn't say we must eat them. Some religions require worshipers to be vegetarians. "

"Name one."

"Jainism and branches of Hinduism and Buddhism. Also, Sichism."

"How long has this bothered you?"

"About a year. Genesis says we should eat vegetables. Here's a quote. 'And God said, 'Behold. I have given you every plant-yielding seed that is on the face of the Earth and every tree with seed in its fruit. You shall have them for food.'"

"We eat vegetables. We raise a garden. We buy vegetables from the farmer's market."

"Those are good things, but they didn't stop Dad from killing Max or you from cooking him."

"Maybe you're going through a phase. Maybe you'll grow out of it."

"This is not a phase. It hurts when Dad kills one of my friends. And when one is served for dinner, I get sick. I take Oscar to McBee Slough and watch the muskrats, frogs, and turtles."

"Your father will be upset if you oppose how he runs the farm."

"I can't help it, Mom. Our animals didn't do anything wrong. They think we love them."

Flora's eyes teared up.

"I'll think about it, and we'll talk again. Maybe we can reduce our meat-eating."

"Eating meat is evil. Richard should have a chance to be a vegetarian. I need you to work on Dad."

"Honey, there isn't much I can do."

"I'll soon be fourteen. Maybe I'll run away."

"Don't say that. I'll think about how I can help you."

"Thanks, Mom. This is important."

***

The next day, a neighbor found Ethan and Richard sowing grass seeds in a field they had cleared of brush. Frank Heesacker sat on the tailgate of his pickup and negotiated.

"I need meat for my freezer. Your goat herd is bigger than you need."

"True."

"Would you sell me eight?"

"If we can come to terms."

"How much per goat?"

"Dead or alive? Do you want me to kill and dress them?"

"Yeah. My wife doesn't want to hear the bleating, smell the crap, and contend with the guts and hides. I want you to dress them."

"Do you want the hides?"

"No, but I want the meat cut and wrapped."

"I need $12 a goat."

"I'll pay $8."

"No, thanks."

"I thought you needed the money. I can buy meat at a grocery store."

"For a price. Maybe you'd like to buy vegetables instead."

"Will you take $9?"

"I need $12. Besides gutting and skinning the goats, I've got to cut and wrap the meat. I'd lose money, what with the time and butcher paper. If you buy them alive, I can let them go for $5 a goat."

"I thought I was doing you a favor."

Frank got down from the tailgate and slammed it shut. He drove away.

***

The next day, Frank found Ethan and Richard in a different field.

"Here's your money. I want eight goats killed, dressed, and packaged."

"Pick up the meat the day after tomorrow. I'll kill the goats tomorrow, and it'll take me a while to process them."

***

Ethan handed the money to Flora.

"We have a problem with Laura."

"Not Laura."

"She decided a goat's life equals a human's. Please give Frank his money back."

"I gave my word. I've got to stick with the deal. Why didn't you tell me about Laura's objection when Frank first showed up? I could've said no."

"I wasn't there when you dealt with Frank. I was here, and you were in a field somewhere. I didn't know someone would come out of the blue and want to buy goats."

"I'll ask Laura to understand our situation. She needs to understand what it takes to raise a family. Maybe this is for the good. She has to grow up sometime."

"Be gentle. You know how sensitive she is."

"But she has to understand that we're a farm family and that we eat meat."

"Include me in your meeting," Flora said. "Maybe I can help her understand what it means to be an adult."

***

That evening, Ethan and Flora met with Laura. She spoke before her dad or mom could. She was looking at Ethan.

"Did Mom tell you of my request? Can we find a way to stop killing our animals?"

"You're going on fourteen. You're a big girl now. So I want you to take this like an adult. We're a farm family. We eat meat. You personally don't have to eat it. Mom and I will never require that. And if Richard follows your example, we'll accept that. But don't expect Mom and me to change our diets."

"I can accept that. I don't have to eat meat. And maybe I can get Richard to copy me. Thanks, Dad."

"Wait, Laura. Let me finish. Do you remember Mr. Heesacker?"

"Yes."

“Just today, he bought eight of our goats. I’ll kill them tomorrow. He’s paying a good price.”

“Can’t you give the money back?”

"I can't back out after giving my word."

“I’ll do extra work that you won’t have to pay me for.”

“I appreciate your offer .... Please understand.”

Laura began to cry. She went to her bedroom.

***

The next day, Ethan used a .22-caliber rifle to shoot eight goats. Richard brought them to his dad one at a time.

Richard pretended the goats were Saladin’s soldiers and that he and Ethan were fighting in the Third Crusade. He pretended they were on the hill called Ayyadieh.

***

Laura didn’t help Flora with the chores. She stayed in her room. After two hours, she went to the corral to see the goats that Ethan had not killed. She called each by name, petted the animal, and cried. She went back into the house and found a packsack. She walked to the slough. Oscar went with her.

When she arrived at the slough, she filled the packsack with rocks. She set it on a stump and backed up to it. She kneeled and worked her arms through the shoulder straps. When she got the pack on, she stood up, swaying under the weight.

"Oscar, you're a forever friend. Go home. Goodbye."

Oscar got between Laura and the water. When she tried to get around him, he moved in front of her. She managed to get past him. When Laura waded into the water, Oscar bit into her shirt and pulled back. But he could not stop his friend.

Laura kept going, wading into deeper water. The loaded packsack weighed her down. She sank below the surface, blacked out, and drowned.

Oscar swam to shore and headed home.

***

Frank met Ethan at the barn. He got the meat and left. It was getting dark. Ethan started for the house. A dog barked.

Ethan bent down to pet Oscar.

"You're wet. What's up?"

He heard Flora call, "Laura, where are you?"

Ethan entered the house.

"Flora, what's wrong?"

"I can't find Laura; I've looked everywhere."

"How long has she been missing?"

"She was holed up in her room. I checked on her two hours ago, and she wasn't there."

"Oscar is wet. Laura might have gone to the slough."

'I'm calling Sheriff Barkley," Flora said.

"Send him to the east shore of McBee Slough near the picnic area. Richard, stay here and obey your mother."

Oscar led the way to where Laura had waded through the rushes.

Two deputies showed up. One wore scuba gear and carried a waterproof flashlight. Oscar showed them where Laura had entered the slough. The scuba diver submerged himself in the murky water. He found Laura and carried her to the shore.

The deputies were unable to revive Laura. They photographed her body and the site as evidence for the medical examiner.

***

It was midnight when Ethan got home.

"You're alone," Flora said.

She started to cry. She put her head on Ethan's shoulder and sobbed.

Richard came out of his room.

"What happened? Where's Laura?"

Ethan said, "She drowned in the slough."

***

In the following weeks, Flora blamed Ethan for not backing out of the Heesacker deal. Ethan blamed himself for being too harsh with Laura. Richard blamed both parents for not recognizing the warning signs of suicide. He had lost a great sister, and he assigned blame.

Chapter 5: Psychoanalysis Begins

Death ends a life, not a relationship.

— Mitch Albom
Tuesdays with Morris, 1977

After Laura died, Flora and Ethan's marriage was filled with sorrow. They needed help, but if they were to spend money on therapy, priority would be given to Richard.

On a Sunday, the Martinson family went on a picnic at the Brownsville Pioneer Park. The park lay next to the Calapooia River, a tributary of the Willamette.

Richard skipped rocks on the water while Ethan and Flora moved food and utensils from the car to a picnic table. The family ate lunch and tried to have a good time. But they couldn't escape memories of Laura.

After lunch, the Martinsons left for home without mixing with anyone at the park. When they got home, Ethan and Flora took their picnic supplies into the house while Richard took Oscar for a walk.

Ethan said, "My boss offered me an advance of my wages if we want to get counseling for Richard. He recommended a psychoanalyst in Albany."

"Let's do it."

***

Ethan made an appointment for Richard with Dr. John Glasser. The Albany clinic had psychoanalysts, psychiatrists, psychologists, and sociologists. Ethan met with Dr. Glasser, who described how he would treat Richard. Each therapy session would last fifty minutes once a week.

Ethan went for a walk while Richard saw the doctor

"How old are you?"

"Ten."

"What do you know about psychoanalysis?"

"Beginning in the 1890s, an Austrian neurologist, Sigmund Freud, developed psychoanalysis as a method of diagnosing and treating psychopathology. He continued to develop the discipline until he died in 1939. Freud's mentor, Josef Breuer, had developed what was known as the talking cure. Freud used the concept to develop psychological theory and therapy.

"According to Freud, psychoanalysis prompts the unconscious mind to reveal its secrets to the conscious mind. The discipline uses dream analysis and free association. Once the memories are uncovered, they can be worked with, using techniques."

"Where did you learn this?"

"The Oregon State University library."

"How did you come to study psychoanalysis?"

“As Dad probably told you, I fell and hurt my head. I thought the injury might have lasting effects, so I studied how mental illness is diagnosed and treated.”

“Will you give psychoanalysis a try? Will you let me help you?”

“Yes. You asked me what I know about psychoanalysis. Should I go on?”

“No. We need to discuss how we’ll work together. You’ll lie on that couch over there. I’ll sit behind you so you won’t see me. This will make you free to react in any way you like to whatever thoughts arise. You won’t worry about how your reaction affects me. You’ll report all thoughts without reservation.

“This is called free association. It frees you up to think any way you like, and I, as the therapist, will know that I’m not accidentally causing you to behave in any particular way. You won’t see my expressions or body language. It’ll prompt you to talk about your life, things that happened to you.”

Richard lay down. The session began.

Chapter 6: A Hypothesis about Meat Eating

> He who is cruel to animals becomes hard also in his dealings with men.
>
> --Immanuel Kant
> *Lectures on* Ethis, 1775-1780

During a psychoanalyst session, Dr. John Glasser asked Richard, "Do you know what a child prodigy is?"

"A child under ten who makes a contribution equal to that of an accomplished adult. There are prodigies in math, science, music, and other fields."

"Has anyone referred to you as a child prodigy?"

"Not that I know of."

"Can you name a child prodigy?"

"Mozart composed music at five. John von Neumann mastered calculus at eight. By the time she was nine, Clara Schumann was a gifted pianist and composer."

"Have you contributed to science, art, or some other field?"

"I have a hypothesis that might save millions of lives if taken seriously. But the subject is taboo because humans think they're superior to all non-human animals. I don't know if anyone is humble enough to research it seriously."

"What is your hypothesis?"

"It came to me when my sister Laura drowned herself. She was intelligent, miles ahead of the rest of our family."

"What is your idea for saving lives?"

"I hypothesize that killing and eating animals leads to a mindset that makes humans more likely to kill other humans. Think about the high murder rate and the high percentage of people who eat animals. It seems there's a connection."

"I've never heard that. I don't believe there's any connection."

"Do you eat beef, chicken, or pork?"

"Well, yes, like most Americans."

"Within a minute of hearing my hypothesis, you rejected it. The possibility that I raise is taboo in America."

"Do you have statistics to support your hypothesis?"

"Most Americans eat meat, and there are many murders in America. It seems reasonable that if you kill animals for food or you have a slaughterhouse worker do it for you, you'll find it easier to kill a human or hire a hitman. Realizing that many Americans have a carnivorous diet may make you feel entitled to kill animals, and then it may not be as hard to take the next step of killing a human."

"You've correlated a common diet with a high rate of murders. But correlation doesn't prove causation."

"But correlation can be a good reason to study an issue. I doubt anyone can prove my hypothesis is wrong. Do you know of anyone studying whether meat-eaters are more likely than vegetarians to murder people?"

"Well, even if they are, that wouldn't prove that killing animals is the reason why a person would kill another human."

"That is not a reason to avoid studying the issue. Some topics are taboo in the US."

"Give me an example."

"Men showing emotion is one. Taking time off work for mental health is another. These topics should not be taboo. Nor should be the idea that a carnivorous diet will make some people more prone to kill humans. Criminologists say that kids who abuse animals are more likely to kill people when they become adults. If that's true, we should consider the possibility that killing animals will make some people more apt to rationalize the killing of humans."

"Tell me more."

"Where should we draw the line? Some people will not hunt deer but think catching fish is a noble pursuit. President Herbert Hoover even said that 'Hours spent fishing are not deducted from the hours God has given us in life'. What's the ethical difference between killing a mammal and killing a fish? Why do we kill and eat cattle and pigs but not cats and dogs? Human rationalization needs to be studied."

"Law enforcement would have taken a stand against eating meat if what you say could be true."

"With the vast majority of Americans being fond of eating meat, few people, including cops, want to be seen as opposing it. Cops love eating meat, just like so many other people. They don't want to be seen as potential murderers. The popularity of eating meat makes it hard to oppose the practice."

"What gave you this idea?"

"Dad killed our animals for us to eat. Laura objected. She was upset when he killed the gander she had named Max. I believe that killing Max made Dad callous enough not to care about Laura's feelings. That attitude is exhibited every day on farms and in slaughterhouses throughout the world. "

"If it's wrong to kill and eat animals, why hasn't someone stopped the practice?"

"Over the centuries, many have tried. Vegetarians have stopped meat-eating in their homes. Let me turn your question around. Why do so many people practice carnivorism?"

"They like the taste of animals."

"Isn't that a flimsy reason for killing billions of animals every year?"

"You tell me why so many people kill and eat animals."

"Because there's money in raising and killing animals. The meat industry tells the public that we need vast amounts of protein and that only meat provides enough protein to keep us healthy. Protein is important, but we don't need as much as the meat industry claims. Also, beans, lentils, alfalfa sprouts, and peas have lots of protein."

"State your hypothesis in simple terms."

"Killing animals for humans to eat is cruel. Albert Schweitzer said so in his book, *Reference for Life*. But he stopped short of saying that carnivorous people ate more likely than vegetarians to kill other humans. I'm saying it now. A carnivorous lifestyle makes it easier for some people to rationalize murder."

"You need to get statistical support."

"I may have to do the research myself. The subject is taboo."

"Do you eat meat?" Dr. Glasser asked.

"Yes, but I'm trying to stop."

Chapter 7: Richard Ends Therapy

> A delusion is something that people believe in despite a total lack of evidence.
>
> -- Richard Darkins
> Evolutionary biologist
> Born in 1941

On the twelfth week, Richard did not lie on the couch when Dr. Glasser asked him to.

"This therapy will go on forever if I let it. This'll be our last session."

"Hey, you're the patient and only ten years old. You can't decide on your own that you're done with therapy. Have you talked with your dad about withdrawing?"

"No, but he's struggling to pay for these sessions. He'll be glad I'm withdrawing."

"Well, I'm the therapist, and I know best. My education and experience outweigh yours."

"I've benefited from your therapy. But I don't think I can benefit any further. Please summarize your findings."

"I'll summarize, but I'll tell your dad that you need more therapy."

"I've upset you, but I believe you can be fair in your remarks."

"I'll be fair. First, you're incredibly intelligent. You are familiar with the concepts of the conscious mind, the unconscious mind, suppressed memories, and dream analysis.

"But you have problems. You sometimes think you're ancient King Richard. At times, you almost persuaded me that you are. You don't seem to have control over the matter. This is why we must continue therapy. Working together, we could eventually eliminate the king from your life."

"I don't know what you're talking about. I don't think I'm a king. I know who King Richard was. But he lived in the twelfth century. This is the twentieth century. I don't think I'm an ancient king."

"Your unconscious mind thinks you're King Richard. We need to bring this to the attention of your conscious mind when it's in a receptive state. Then, you can combat the belief. Several times during our sessions, you admitted to thinking you're the king, but the realization slipped away from you later. We made progress but then lost the ground we had gained."

"Maybe I have momentary memory problems, but since I'm not tormented, why should I bankrupt my parents by continuing with therapy? Thank you for all you've done."

Dr. Glasser told Ethan of Richard's need for continued therapy, but Ethan agreed with his son. Ethan said he would bring Richard back if the need arose.

Chapter 8: Developing Credentials

Money grows on the tree of persistence.

— Japanese proverb

Years went by. It was 1985. Richard seemed to have recovered from his fall.

The Martinsons and the Jewels were pleased when Oregon State University accepted their sons. Richard and Bryce each received a small scholarship that was awarded when they graduated from Central Linn High School. They shared a dormitory room at OSU.

It was their second year at the university. They were studying far into the night, as was their custom.

Richard said, "You need to help me publish papers so we 'll be hired into professional positions."

"I'm doing my part."

"You took too long to research Valium compared to Librium. I published a paper on my own while you dawdled."

"Dawdled? I worked as fast as I could. I had trouble getting information. I'm ready to submit the paper."

It was challenging to publish scientific papers while meeting requirements such as attending classes, giving classroom presentations, and taking exams. But Richard and Bryce agreed there was no better way to distinguish themselves. To hold a student office, such as class president, was impressive. Becoming an office holder in local government, such as a school board member or city council member, could impress prospective employers. However, published scientific papers were the most relevant to the career that a university graduate might hold. So, Richard and Bryce concentrated on researching and writing.

When the two men received their bachelor's degrees, they were tempted to attend another university for their postgraduate education. But, after considering their options, they decided to stay at OSU. Eventually, they graduated with advanced degrees. Richard got a PhD in psychiatry, graduating at the top of his class. Bryce earned a PhD in pharmacology with a focus on psychotropic drugs.

Both men had submitted employment applications during the six months leading up to their graduation. Richard landed a career position with Green State Medical College in River Town. The college hired him to help revive its psychiatric department. It was a challenge for a person's first career position, but because of his publications, Richard's credentials were stellar.

After accepting the job, Richard learned that the school's psychiatric department was also seeking a pharmacologist. Richard recommended Bryce for the position. Bryce was hired.

Richard bought a house. Bryce rented from him.

Chapter 9: Honoring the Contract

Creditors have better memories than debtors.

--Benjamin Franklin
A founding father of the United States

The medical college employed Richard and Bryce, but their offices were in the nearby Veterans Administration Hospital. The clinical trials were paid for by the pharmaceutical companies that owned the drugs being tested.

The hospital's chief financial officer, Kenneth Shaw, told Richard of a problem between the hospital and the medical college.

Shaw said, "You have a fresh perspective as the new manager of clinical drug trials. As trials are done, the college is not providing the hospital with its share of the money that the two agencies agreed on. Will you help me settle this issue?"

"Bills must be paid promptly if the medical college and the VA are to have a successful partnership."

"Will you help me get the college to pay up and stay current?"

"I've seen the bills. This is an old problem. I'll help you."

***

Shaw made an appointment with Arnold Picket, the college's chief financial officer. Richard went with Shaw to the meeting.

Shaw said to Picket, "Richard is your school's employee and my hospital's tenant. As you know, the college hired him to oversee the clinical drug trials done at the hospital. Despite our sharing agreement, your payments have been haphazard. I hope he can help us settle this dispute. I won't wait any longer for the college to pay for using our hospital, equipment, and staff. I'll elevate this issue to top management."

Picket said, "The college paid you less because fewer drugs were tested while we replaced nonperforming employees with Richard and his friend, Bryce Jewel. When drug tests are delayed, it follows that the money the college pays to the hospital also is delayed."

"That accounts for part of the problem, but you still need to pay rent every month. Even if you did no testing in a given month, the VA still provides a hospital ward, a laboratory, a nurse's station, and offices for your managers."

"You haven't been getting rent money from us?"

"Not on a regular basis."

"I didn't realize that. I'll fix the problem today. Future drug testing income will depend on the contracts Richard establishes with drug companies. The fewer the contracts, the less the drug money."

"I understand," Shaw said. "But Richard will get contracts, and I don't want the college to fail to pay us what we're entitled to."

"You brought my college's new employee here to pressure me. Richard, do you know why you're here?"

"The college hired me to manage the agreement between the college and the hospital. I'll see that we do reliable drug testing."

"You work for the college, not for Ken's hospital. Don't forget that."

Richard said, "Do you have advice on how to settle this dispute?"

"Create a nonprofit corporation. Do this by applying for a federal employer identification number for the drug-testing ward at the hospital. Come up with a business name. Ask the state to identify the ward as a nonprofit business entity. This will cause the state to create a corporation. Then you need to make that new corporation disperse the pharmaceutical payments according to the agreement between the college and the VA."

"Anything else?" Richard said.

"Give the VA ten percent of the funds that the college will receive from each pharmaceutical contract."

Shaw stood up.

"Thirty-three percent. If we don't get a third, you can't use our facilities."

"Ha, ha, ha. I rattled you. You'll get a third of the proceeds. I'll sign the papers. Good-bye."

Ken and Richard began walking back to the hospital.

Ken said, "I don't trust Picket. He'll be a problem for you. He's always wondering what's in it for him."

"That's normal."

"Wait until he does it to you."

“I’d report it to higher-ups.”

“His bosses won’t correct him. He knows too much about the college’s questionable financial practices and which officials are complicit.”

Chapter 10: Forming a Partnership

> The first rule in keeping secrets is nothing on paper. Paper can be lost or stolen or inherited by the wrong people. If you want to keep a secret, don't write it down.
>
> — Thomas Powers
> *The Man Who Kept the Secrets, Richard Helm and the CIA*, 1979

Arnold Picket oversaw the medical college's budget. His approval was required for any major undertaking proposed by any college employee.

Arnold was president of the River Town Lions Club. Richard joined the club so he could get to know Arnold better. Soon, they were on a first-name basis. The club's annual projects included a yard sale and free dinners at Thanksgiving and Christmas.

In February 1988, Richard saw Arnold post a for sale sign on a bulletin board in the Student Union. Arnold was selling a Buick Somerset.

Richard phoned the number on the ad and pretended not to know who the seller was. Arnold's wife answered the phone and invited Richard to see the car.

Richard pulled up to the curb. Arnold came out of the house.

"Arnold, I had no idea that you're the seller."

"I bought the car new. I hate to part with it."

Richard pretended to check the Buick's mechanical condition. He took the car for a test drive. He and Arnold agreed on a price. Richard wrote a check.

Arnold invited Richard into the backyard patio to drink a beer. After three stubbies each, they talked freely.

"You must be a good investor to afford this home. "

"I have my military annuity, and I began investing while I was young."

"A mansion with acreage on a cul-de-sac."

"Are you finding it hard to live on your salary?"

"I've thought of a way to increase it."

"Does this involve me?"

"It could," Richard said. "Can we talk off the record? Engage in what if?"

"Sure. Off the record. That's when we get creative."

"Many people have learned how to make their jobs pay extra. That's what I want to do. To make my plan work, I need the help of someone with clout at the college."

"You spent months doing things like Lions Club projects and buying this car. You want something from me. So I had you investigated."

"Oh?"

"I hired a private detective to check you out. I realized that you wanted to buy the Buick to get a chance to talk with me in private."

"You knew I saw you post the for sale sign at the Student Union?"

"Yep. I had seen that you were at the Student Union most afternoons. I posted the sign when you were there. Recognizing how motivated you are, I added sixty percent to the price. No one else called about the car because the price was too high."

Both men laughed.

"Did you reach any conclusions about what I want?"

"You want something that I can provide as the college's chief financial officer. And it has to do with clinical drug trials because that's what you do. Go ahead and propose a business deal."

"You told me how I could pay the VA its share of the drug trial money. You said I should create a nonprofit corporation and have it distribute the drug company money between the college and the VA. I did as you said, and it worked perfectly.

"To understand the rest of what I'm going to say, you have to know that clinical drug trials depend on institutional review boards."

Arnold said, "I understand that IRBs have to approve the plan that a researcher proposes. Without IRB approval, the researcher legally can't do the drug trial."

"A staff member and I set up a nonprofit corporation of our own. We named it Pharm Tech. We avoided using any name that was likely to be associated with the medical college. I'll be able to use commercial Institutional review boards, which are more lenient than the nonprofit one on campus. By using commercial IRBs, I'll get faster approval of my proposed drug tests."

"Who is the staff member in the corporation with you?"

"Bryce Jewel, a psychopharmacologist. We grew up together. The college hired him on my recommendation."

"I remember the name. Do you trust him?"

"Completely."

Arnold asked questions about Bryce. Eventually, he said, "I can live with Dr. Jewel being involved. You grew up with him. And you need help running the drug trials. But share as little as possible with him. And don't take anyone else into your confidence. If you bring in even one more person, I'll back out of our agreement. We must keep this venture to ourselves."

"I agree. Thank you for advising me on setting up a nonprofit corporation. Right away, I saw the value of controlling such an organization."

"I don't understand," Arnold said.

"That's because I haven't given you the rest of the story."

"Fill me in."

"Few people want to serve as test subjects. Most contract research organizations take months and months to recruit enough test subjects for a clinical trial. A drug can't be marketed until the FDA approves it. Drug companies claim that every day spent preparing to run a test results in a loss of $1 million in sales. I can get lots of contracts to test drugs if I can do them quickly. Drug companies speed up clinical trials by paying research contractors $20,000 or more per test subject."

Arnold sat up straight. His eyes lit up.

"How many subjects do you have in a typical clinical drug trial?"

"You need one hundred subjects in an experimental group and one hundred in the control group. But, you may want to test more than one drug, so you might need an additional hundred subjects for each additional drug you test. And you need healthy volunteers and a placebo group. Pretty soon, you're talking three hundred subjects for an average drug trial. I avoid overestimating by figuring that I'll need at least two-hundred-fifty subjects for most any trial."

"Even if we use a figure of two hundred fifty subjects in a trial, at $20,000 a subject, that's half a million dollars. Can you get approval faster than other contract research organizations?"

"Drug companies are not only desperate for test subjects; they're awash in money. So, they're happy to pay for quick trials. Most of the time is not spent on testing but on recruiting enough subjects to do the tests.

"Other research contractors don't seem to recognize how desperate drug companies are. I achieve speedy trials because I pay twice as much as other contractors do for each test subject. But I don't exceed my budget because I get the money from the drug companies. How? By persistently asking for the money. I recognize the frustration and the wealth of the companies, and I drive a hard bargain."

"But other contract companies aren't dumb. Competition will increase."

"But I can tie up a lot of markets first. Especially with a technique that Bryce and I developed. I don't think any other contractor is using this method."

Arnold said, "Describe it."

"It isn't something to be proud of, but it's lucrative."

"All's fair in love and war. If we're going to be partners, I need to know your methods."

"Most of the test subjects with schizophrenia are young men. We hire good-looking women to recruit men for testing. It works beautifully."

"How far do you take it?"

"Far enough to get the test subject enrolled and to get the fee from the drug company."

"That's prostitution."

"As far as I know, flirting is as far as it goes. Neither the women nor the men admit to having sex."

"Your phrase 'as far as I know' suggests that if necessary, you'll be prepared to admit that you're ignorant of what's going on."

"That's right."

"How do you know if the female recruiter and the male prospective test subject are being honest? You aren't there every time they meet. And I doubt you're filming the meetings."

"We use the honor system."

"Well, as long as the recruiter and the test subject get something out of the deal and as long as you don't tell your lovely recruiter specifically what to do, you probably won't get in trouble."

"The college has a bank account in which I'll deposit a small number of drug company payments. Whatever drug company money the college gets has to come through me. The college has no other way of knowing how many trials we do.

"The money we get for clinical trials comes from drug companies. The funds are not from tuition or grants, which are the college's main sources of income and which are closely monitored. Everything about tuition and grants is transparent. But the school is clueless about the money that comes from clinical drug trials."

"Someone above you, like the college president or a financial officer, could check your bank account."

"That's true. But that's the college's existing bank account that you're talking about. Checking into it won't tell the inquirer much except that the account isn't very active."

Arnold said, "But you concede that college officials can get a court order to search for bank accounts."

"That's right. So I won't give them cause to become suspicious. I'll set up a new bank account for our drug trial business. I'll give it a secret account name. Only you and I and the bank will even know the account exists."

"If you know what name you'll use, tell me now."

"Why?" Richard asked.

"To show that you trust me, that you mean what you say."

"I'm leaning toward Fairway Industries."

"I like the name because it's vague. It doesn't suggest that the account holds the fees you will charge drug companies."

Richard said, "Of course anyone getting into the account will see immediately what we are doing but at least the name will not broadcast what kind of business we are running."

"Your idea is to divert money to ourselves so the college never sees it."

"That's right. We'll bypass the college's safeguards. Administrators would have no reason to be suspicious unless they were tracking the drug trials through some other means."

Arnold said, "The college hasn't set up any such means."

"That's the point. Unless someone were to get a court order, no one but I would see the new bank account."

"Is there more to your plan?"

"You'll need to be alert. You'll be our first line of defense. Everything financial at the college goes through you."

"You want me to defend you if someone gets suspicious about where pharmaceutical payments are going."

"That's right. Will you throw in with me?"

"I want half the proceeds. College faculty and staff might catch onto your scheme. You need me to quell their suspicions."

"Fifty percent is too much."

"Take it or leave it."

"I'll take it."

"Get together figures on the potential income. Don't put anything on paper. Keep this between the two of us."

"That's the plan. I'll come back tomorrow to get the Buick."

"Just one more thing," Arnold said.

"Yes?"

"It would be a fatal mistake to say a word of this to anyone."

Chapter 11: Suite Seven

Money never made a man happy, nor will it.The more a man has, the more he wants.

— Benjamin Franklin
A founding father of the USA
1705-1790

Two days after Richard bought Arnold's car, they met on the medical college campus. They sat on a park bench.

Arnold said, "How important is money to you?"

"I need money to feel secure."

"Was your family poor?"

"Dad works for a grass seed farmer. My parents still live in one of his shacks. My sister died at thirteen. I had a rough life."

"Have you thought about who you'll harm with your scheme? If you hurt the wrong people, they'll fight back."

"The only victim will be the college, which will scarcely notice the loss. It doesn't bring in much money because of poor management. The college didn't build the profitable business that you and I are about to create. It won't lose money that it never had. What it'll lose is the potential for making money. But the money we'll get wouldn't exist without our hard work. We see an opportunity that the college doesn't. We won't harm any individuals. The money will come from rich drug companies, and they'll get the speedy trials that they want."

Arnold said, “What about the test subjects? Will their health be compromised? “

“We’ll protect everyone’s health.”

“Whether your scheme works will depend on your skill.”

“I’m confident of my ability.”

Arnold said, “I looked into the on-campus IRB. People say it’s lax. It doesn’t directly oversee research projects. Instead, it requires only that researchers submit quarterly reports. Additionally, the on-campus IRB has clients beyond Pharm Tech. This means that it does not depend primarily on your business. So, if you don’t use it much, it probably won’t miss your business. This IRB probably will never catch you as long as you use it now and then.”

“I’ll do that,” Richard said.

“Your plan is to use a commercial IRB primarily so you can get approval for drug trials that the nonprofit IRB might reject. I think this plan will work as long as you don’t get too greedy.”

“I’ll be careful and will keep you informed. I’ll tell you how many drug trials I’m doing, how many test subjects I have in each trial, and what expenses I want the college to cover.

“The more expenses we can pass onto the college, the more money you and I will have. But I agree with you that if we get too greedy, we’ll increase our chances of being discovered. I’ll run every major decision past you. You’ll have control over any risks we take.”

“That’s what I wanted to hear.”

***

Richard and Bryce were in Richard's office in the VA hospital.

Richard said, "We do the clinical trials in the hospital. But we need a separate clinic for interviewing test subjects, giving acuity tests, and discussing sensitive medical issues."

"I agree. We need to share some topics with the hospital, but some we must keep secret. I don't want a hospital employee to walk in when I'm asking a drug company to pay a little extra for better results."

Richard said, "Find a clinic for us where we'll have privacy."

***

Bryce found a building a mile east of the campus. A sign said, "Suite Seven." Richard named this clinic "Pharm Tech." It would be responsible for work at both the VA hospital and Suite Seven.

After clinical testing began, it was hard to keep information secret. So, drug companies got a patent for a new drug before they began testing it.

The FDA prohibited the sale of drugs during the testing period. Some of the valuable patent period was used up while no funds were coming in; thus, companies completed testing as quickly as possible.

Drug companies hurried to find human subjects. Scarcity of test subjects was the biggest cause of the delay in getting new drugs to the market. Companies competed for early dates for their clinical trials.

Pharm Tech solicited test subjects in messages disguised as public service announcements. The clinic solicited advertisements on television, radio, and in newspapers. It distributed individual mailings, posters, and flyers throughout the River Town area.

Chapter 12

The world is not fair. If you persist in presuming it is, you will create a lot of misery for yourself.

—Fred Green
American professional baseball player
1933-1996

Richard and Bryce were in Richard's office in Suite Seven. They were working out the details of their drug testing program.

Richard said, "Beautiful women are the best recruiters of men with schizophrenia. But there's a catch. Do you know about the EEO?"

"Equal Employment Opportunity. It's a federal program to prevent discrimination in the workplace. The EEO Commission could prevent us from hiring good-looking recruiters. Our classified ad can't say that ugly people should not apply."

Both men laughed.

Richard said, "The job applicant who looks best on paper might be homely enough to discourage men from applying for a test position."

Bryce ran a classified ad in the *River Town Times* for jobs with Pharm Tech. Many women and a few men came to Suite Seven with resumes in hand. Three female clinical employees — Loretta Getz, Susan Goodwin, and Rosemary Adams — conducted the interviews in the waiting rooms. Applicants filled the clinic.

Bryce showed up wearing a dress and a wig.

Loretta said, “What’re you doing?”

“I don’t want anyone to know that I’m involved with the hiring. When someone goes to work for me, I want her to think I’m neutral.”

“You could pretend to be a male applicant.”

“Someone would recognize me.”

As part of his disguise, Bryce held a file folder. He gave Loretta instructions that she passed on to Susan and Rosemary.

Bruce stared at a heavyset female applicant.

“Have Susan send her away.”

An elderly woman entered the waiting room.

“Have Susan get rid of her.”

The next applicant was good-looking and had long, black hair.

“Hire that one.”

After four hours, Bryce gave Loretta final instructions.

Hire the good-looking brunette and any two of the five that I okayed.”

***

Rosemary Adams interviewed Sandra Porter.

"Miss Porter, before you enrolled at OSU, you were an airline stewardess. Did that job help qualify you for our open position of clinical trial coordinator?"

"Your classified ad called for applicants with the ability to relate to sick people and to schedule their medications. As an airline stewardess, I saw to the comfort of hundreds of people, sometimes in turbulent skies. Occasionally, passengers suffered motion sickness. I can relate to ill persons individually or in groups."

"You would have to do medical tasks like draw blood and understand medical charts."

Sandra smoothed her hair back from her eyes.

"You asked about my experience as an airline stewardess, but I also completed two years of study at OSU."

"A lot of the test subjects will be young men with a mental illness. Most are gentlemen. But what would you do if that rare exception were to attack you?"

"Since I reached adolescence, males have tried to have their way. My mother taught me how to fend them off. As part of my training as a stewardess, I took karate courses. I'm not worried about working with men regardless of their mental condition."

Pharm Tech hired Sandra Porter, Cindy McMullen, and Betty Schroder and called them project coordinators.

Chapter 12: A Flawed Memory Test

> Reference for life affords me of my fundamental principle of morality, namely that good consists of maintaining, assisting, and enhancing life, and to destroy, to harm, and to hinder life is evil.
>
> --Albert Schweitzer
> *Civilization and Ethics,* 1924

Richard wondered why Laura had felt so strongly about whether goats lived or died. He liked goats but would not have taken his life because they lost theirs. He wondered what philosophers had said over the centuries about the lives of nonhuman animals.

Richard read that Albert Schweitzer developed an ethical philosophy that he called "Reference for Life," winning the Nobel Peace Prize in 1952. Schweitzer's writings helped Richard understand why Laura had felt so strongly about the deaths of a gander and goats. They helped him understand what she must have felt.

Richard did not know of any living person who could have influenced Laura's decision to not eat animal products. She seemed to have on her own become aware of what the literature called ethical veganism. She supported animal rights due to her independent thinking.

Periodically, Richard thought of becoming a vegan. But the idea vanished when his personality changed to that of King Richard.

***

Kimberly Bissell was intrigued by an experimental drug for treating Alzheimer's. She brought her husband, Adrian, to Suite Seven.

The retired teacher couldn't remember which state he lived in. Coordinator Susan Goodwin took him to a back room. She gave him a mental acuity test, which had a scale ranging from zero to thirty. A score of 10 was required to enroll in the upcoming study.

Susan and Adrian returned to the lobby.

Susan told Kimberly, "He scored eight, but I gave him twelve."

Three months into the Pharm Tech study, Kimberly realized that her husband wasn't improving. She took him to a neurologist not associated with Pharm Tech.

"Since Pharm Tech's acuity test, has Adrian improved, stayed about the same, or worsened?"

"He hasn't changed."

The neurologist gave the same test to Adrian that Susan had given him. His score was two.

Chapter 13: We have informed consent because of Nazi experiments.

> The most complete and authoritative statement of the law of informed consent to human experimentation is the Nuremberg Code.
>
> --George J. Annas, et al. *Informed Consent in Human Experimentation: The Subjext's Delimma*, 1977

Men diagnosed as having schizophrenia lived in the Whitmore group home in River Town. Some suffered from hallucinations and delusions. Some were in medical stupors. A few were doing well.

Sandra, Cindy, and Betty went to the group home to recruit subjects for a Pharm Tech drug trial. They were in the day room.

Billy Nelson was assembling a jigsaw puzzle. Sandra introduced herself.

"I'm pleased to meet you, Miss Porter."

"Call me Sandra."

"Sandra, why did you come to see me?"

"My medical clinic believes you have schizophrenia. Pharm Tech is developing a drug to treat the disorder."

"How did your clinic make that decision? By law, my medical records are confidential."

"Pharm Tech knows that people diagnosed with schizophrenia live here. I saw you sitting here and decided to talk with you. I hope I'm not offending you."

"That's all right, Sandra. You seem like a nice person."

"You seem like a nice person, too."

"I don't have schizophrenia. Does that take away your reason for talking with me?"

"No, you could participate in a trial by serving as a healthy volunteer."

"You don't believe I'm healthy. You see me as a guinea pig."

"If you participate in a trial, Pharm Tech will pay you $150 a day plus meals and lodging. Your participation might lead to the development of a wonder drug. You'd help other people who have schizophrenia."

"That's important. If I can help other people by being in a study, I'll be happy to do it."

"Helping others is important to you."

"I don't want to have lived in vain."

"That's good of you."

"Will you do the study, Sandra?"

"I'd be one of the technicians to give you pills."

"Are you sure that your boss would assign you to my study group?"

"I'll ask him to."

"I've heard of people getting sick. Some died."

"We'll minimize the risk. I'll ask Pharm Tech to pay you $175 a day and cigarettes."

"I want a carton of Camels before the trial begins. And I want $200 a day. I want to see you every day."

"I can approve the Camels and the $200. But I don't know what work I'll be assigned. So, can we agree that I'll see you when I can?"

"I want you to be my girlfriend. You said I'm a nice person."

"But I can't promise that I'll see you every day. Can we agree it'll be me or another girl?"

"Seeing you every day would make me feel safe."

"Other people will be there watching out for you."

"You think you can talk me into anything. I'm not going to do it."

"Maybe you'll change your mind. Can I come back to see you?"

"I'd like that. How about next week?"

"Let me check my calendar."

***

At another table, Cindy McMullen was talking with Frank Lester.

Frank said, "Pharm Tech and the drug company would be the big winners. They're not merely being good citizens when they test drugs. They get a lot out of the deal.

"But you get things out of the deal too."

"You offered me $150 a day. How much does Pharm Tech get from the drug company for each person tested?"

"We're talking about whether a clinical test would be right for you. It doesn't matter what Pharm Tech gets."

"I shouldn't have to risk my life because your clinic wants to test drugs. Tests on human subjects are not supposed to be done unless there is a great need. The Nuremberg Code, the Helsinki Declaration, and the Belmont Report all say this. The potential benefits must outweigh the risks. It's insufficient to merely want to improve a drug's marketing status."

"How did you learn that?"

"I had uncles who were identical twins. They died as children in 1944 in the Auschwitz concentration camp in Poland. They were part of a genetics experiment done by Josef Mengele, a Nazi doctor nicknamed 'The Angel of Death.' Years later, when I learned of the deaths of my uncles, I read the Nuremberg Code. A few years after that, I read the Helsinki Declaration and the Belmont Report. Before asking anyone to participate in drug trials, you should learn what Hitler's doctors did when they experimented on prisoners of war during the Holocaust."

"You're saying that today's clinical trials are related to the Holocaust. I've never heard that."

"You need to study the cruelty of Josef Mengele, Karl Brandt, and other Nazi doctors before you enroll patients for your experiments. The horrors of World War II concentration camps are the main reason why restrictions are placed on clinical trials. Read the Nuremberg Code."

"Never heard of it."

"Nazi experiments during World War II were heinous. Suffering and death were prevalent. Surgery was done without anesthesia. Experiments were done on sterilization, counteracting chemical weapons, and the treatment of war wounds. Experiments were done on vaccines and drugs and the survival of harsh conditions."

"What kind of harsh conditions are you talking about?"

"An example would be a plane crashing into the Arctic Ocean. For the doctors to learn how long a pilot might live in freezing water meant that test subjects were to be frozen to death in a Nazi lab."

"Oh."

"The Nuremberg Code was formulated in 1947 by Americans who judged Nazi doctors accused of doing cruel experiments in the concentration camps. The collective judgment led to the establishment of research ethics for experimentation on humans."

Cindy said, "I'll do my homework. But for now, I want to make sure you benefit from taking part in a drug test. You'll get money, meals, lodging, and cigarettes."

"You changed the subject."

"I apologize. Pharm Tech sent me here to sign up people for drug trials. I'm learning from you, but I need to go back to the clinic with at least one enrollee. You'd have a chance to help others."

"My participation might advance scientific knowledge if the drug company publishes a report on the trial. But if the experiment doesn't make your company's drug look good, a report will never be written."

"It's amazing what you know."

"You're impressed because you assume that someone diagnosed as having schizophrenia is dumb. Schizophrenia screening doesn't measure IQ. You think I'm too dumb to see that you're trying to exploit me with your good looks."

"I'm making no such assumption. Pharm Tech sent me here to inform people like you about the benefits of participating in a clinical trial. "

"I hear discrimination when you say 'people like you'."

"Ouch. Let's start over. Tell me about yourself so I won't make stupid assumptions."

"Suffering psychosis once in a while doesn't wipe out my ability to reason the rest of the time."

"Are you here voluntarily?"

"I'm here by court order. If I were to consent to your drug trial, a court would have to approve my participation. You don't seem to know much about how the legal system regulates mental illness."

"You're teaching me. What did you do for a living before you came here?"

"I worked for the Oregon Bureau of Labor. I tracked unemployment rates."

"You're smart. Are you married?"

"Do you have something in mind?"

Both laughed.

Cindy said, "Well, I am attracted to you."

"If I entered the drug trial, would I see you often?"

"I could arrange that. You haven't asked if I'm married."

Frank said, "I don't see a wedding ring."

"I was engaged, but that relationship ended long ago."

"Tell me more about the drug trial."

***

Betty Schroder was 20 feet away, talking with Dino Scaletti.

"Thank you for making time for me, Dino."

"I'm happy to have a visitor."

"How long have you been here?"

"A year. Most of the men stay only a few weeks. I think they lost me in the tracking system."

"How long has it been since a doctor saw you?"

"Nine months."

"Are you taking medicine?"

"Not since my doctor stopped seeing me."

"Are you familiar with how clinical trials work?"

"I was in one."

"How did it work out?"

"I got sick, but I recovered. I got paid."

"Are you willing to enter another one? You'd be in the VA hospital. Green State Medical College would do the test."

"I'll do it if you can get my doctor's approval. He probably won't care. I don't know if you'll be able to find him."

"I'm glad that you're willing."

"There's nothing to do here. I'm bored."

Chapter 14: The Hero I Needed

> What is being done to us and why? Whatever is being done has made pharmaceutical companies, which barely existed before World War II, into giant corporations and the darlings of Wall Street. It has created a medico-pharmaceutical complex that appears to have gradually shifted from discovering treatment for major diseases to medicating aspects of the human condition.
>
> --David Healy, MD
> *The Creation of Psychpharmacology*, 2002

After lunch on an April day in 1988, Ira Cohen and Tom Vitchner returned to their dormitory room. They were classmates in psychiatry at Oregon State University.

Tom said, "You wouldn't let me buy you lunch. Noodle soup was all you ate. You can't keep on that way."

"I'll get on my feet. I'm taking a job."

"Part-time?"

"Full time. I'll drop out of school for a few terms."

"You'll delay your medical practice."

"But I won't go bankrupt before I establish one. And I don't want to take out a student loan."

"What's the job?"

"Green State Medical College and the VA Hospital have teamed up to test psychiatric drugs for FDA approval. The college manages a ward in the VA hospital. I'll work for the college."

"What'll you do?"

"I'll help doctors and nurses run the drug trials. I'll make beds, clean up puke, and do laundry."

"How did you find out about the job?"

"Sandra told me. She got a job there herself. She recruits people to be tested and gets a bonus for every person she enrolls."

"Where does she find people to test?"

"Assisted living facilities and shelters for the homeless. Places where people have disorders but little or no health insurance."

***

Ira lacked a role model when he began to pursue a psychiatric career. No one in his family had gone to college. His brother had been diagnosed as having schizophrenia and, at seventeen, died by suicide. Ira wanted to help people who had mental illness, but he had no one to show him the way. Then, he learned of Dr. Richard Martinson.

***

Ira and Ton drove to Portland to attend a symposium on mental disorders at the Oregon Health & Science University.

They examined medical displays in the lobby of the convention center. The two men latched onto briefcases, books, and coffee mugs. They found seats in the auditorium and settled in to enjoy the seminar.

The speakers included psychiatrists, psychologists, sociologists, and occupational therapists. Ira and Tom had stayed up late the night before, and the drive from Corvallis took its toll. Either Ira or Tom would doze off; his companion would wake him and then fall asleep himself.

Tom said, "We need to go for a walk and wake up."

"Wait. The next speaker is the head of the psychiatric department that I'll be working for."

Dr. Richard Martinson came to the lectern. He spoke without using notes. He described the clinical drug trials that were used to assess the safety and effectiveness of drugs proposed for the marketplace. He traced the history of randomized, controlled trials. He described the difficulties that researchers had dealt with in early clinical trials. He told of the creation of rating scales and how they enabled clinicians to come up with numeric scores for the trials. He said that a randomized clinical trial was the best way to determine whether a drug is safe and effective.

Next, Dr. Martinson participated in a panel discussion on why psychiatry is a branch of medicine. He gave a talk and then debated other panelists.

Tom said, "Martinson is the best speaker. You'll learn a lot from him."

"Maybe I've found the hero I've been looking for."

The next week, Ira reported for work, joining Dr. Martinson's team.

Chapter 15: Suicide Attempt

Honesty stands at the gate and knocks, and bribery enters in.

--Barnaby Rich
English author and soldier
1540-1617

Pharm Tech was up and running. Now, it was growth that was needed. Richard and Bryce were making plans at Suite Seven.

Richard said, "We need to tell the drug companies that we can get their drugs approved faster than any other contract research organization."

"What do you want me to do?"

"Write to the major drug companies. Use the college's stationery."

"This is upsetting."

"What's wrong?"

"It just hit me – what we're about to do," Bryce said.

"We're just promoting our business. Are you getting cold feet?"

"Don't misunderstand me. I'm with you all the way."

"Stop worrying. Enclose our resumes and a list of our publications."

Pharm Tech mailed letters to twelve pharmaceutical companies. Responses came from Abbott Laboratories, Bristol-Myers Squibb, Johnson & Johnson, SmithKline, and Eli Lilly. These companies came to meet with Richard and Bryce. They inspected the VA Hospital and Suite Seven. All contracted for drug trials.

***

Bryce gave Cindy McMullen a bonus for every subject she recruited. To recruit as many subjects as possible, Cindy checked out the assisted living facilities in River Town. She went to homeless shelters and asked friends and relatives for leads. She found a prospect in Samuel Winegart.

Cindy interviewed Sam. He agreed to enter a test of Fremontria for treating depression. He had been depressed, but no longer had symptoms.

Cindy reported to Bryce in Suite Seven.

"I found a 65-year-old man willing to participate in our Fremontria trial. But he hasn't had symptoms for two years. What should I do?"

"Healthcare providers used to think depression was episodic. Now we know it's chronic. He's still sick. Sign him up."

Sam moved into the VA hospital. Within a day of his first dose, he was disoriented.

On the second day of the drug trial, Ira entered a restroom in the hospital's drug-testing ward. Sam was hanging from a shower rod. He had used bath towels and had placed his weight on the noose by bending his knees. He was strangling.

Ira wrapped his arms around Sam's waist and lifted him.

"Help. Suicide attempt. Help."

Sandra hurried in and untied the towels. Ira carried Sam to a gurney and rushed him to the emergency department.

An hour later, an emergency nurse told Ira and Sandra that Sam had survived. They congratulated each other.

Ira said, "My shift is over. I'll see you tomorrow."

Sandra reported Sam's suicide attempt to Richard and Bryce. They were in Richard's office in the hospital. Richard stared at his computer screen.

"I've broken the blind. Sam was taking Fremontria. But let's not jump to conclusions. There might have been a cause other than the drug that was being tested."

Sandra said, "I assume you'll report this to the FDA and the Hancox Company. If you don't, I will."

"That wouldn't help," Richard said. "If we involve the FDA, our clinic could be shut down. We can't help people if we're not in business."

"I'm not telling you to shut down, just to let authorities evaluate what happened to see if we can improve."

Bryce said, "We'll take care of Sam at no expense to him. We'll do the right thing without being forced."

"Your responsibility extends beyond Sam."

"Don't contact the FDA," Richard said. "A regulatory investigation won't help. We'll pay you to keep this quiet."

"I want to see Sam before I decide."

"Bryce, take Sandra to see Sam."

They went to Sam's hospital room. He was sitting up in bed.

"It was a blur. I'm better now."

"It's not your fault, Sam," Sandra said.

"Sometimes bad things just happen," Bryce said. "Pharm Tech will see that you get well."

"If I were you, I'd get a lawyer," Sandra said.

Bryce said, "Richard is waiting for us."

Sandra and Bryce returned to Richard's office.

"Sam isn't holding a grudge, but I'm still upset."

Richard said, "Don't tell anyone. We'll compensate you."

Sandra went to the break room. Ten minutes later, she took a phone call. She went to Bryce's office. He paid her $5,000 in one-hundred-dollar bills.

Chapter 16: Keep Trying

> Perseverance is not a long race; it is many short races, one after the other.
>
> — Walter Elliot
> Scottish politician, 1888-1958

Richard and Bryce were in Richard's office at Suite Seven. Richard said, "We need a commercial, institutional review board that'll help us reach our goals, that won't drop in on us without warning, and that'll allow us to report on paper only. We need one that'll quickly approve our proposed drug trials. Before I hire an IRB, I want it to signal a willingness to approve whatever we propose."

"It'll cost us."

"But we'll end up making a lot more money. We'll forge ahead of all the other drug-testing companies."

Bryce interviewed commercial committees in River Town. None seemed lenient enough. He broadened his search to other cities.

Bryce met with the director of a commercial ethics committee in Salem.

The director said, "Two years ago, a subject in one of our trials reported us to the Office for Protection from Research Risks. The agency eventually decided that we hadn't done anything seriously wrong, but the process we had to go through took time and paperwork."

The next day, Bryce reported to Richard.

"Every commercial, ethical company has been cautious. They might be willing to bend the rules for us, but won't say so. We might have to take a chance on a company and see how it works out."

"That could be a long process. I want a company that reveals its willingness from the start."

"What should I do?"

"Keep going. Sooner or later, you'll find someone who'll let you know that his company is the one we want. The person may have to get to know you before confiding in you."

"Unless a company has a great need for money, I doubt a manager will indicate he's willing to bend the rules."

"Oh, you never know. Someone might have money but want more."

***

A week later, Bryce interviewed Ted Domain, a manager of a commercial ethics committee in Salem. Ted seemed willing to break the rules. Bryce bought him lunch.

"Pharm Tech will pay extra to a helpful ethics committee. We intend to do many clinical trials."

"My company, Pinnacle Ethics, started up just a week ago. We're looking for clients."

"Pharm Tech has to sign up test subjects very quickly. That's the image we're building our reputation on."

"Come to my office Monday morning."

When Monday came, Bryce was at Ted's office.

Ted said, "I'm working to get Pinnacle off to a fast start. I'll provide the fastest service possible."

"People are afraid that drug trials will sicken or kill them. And there's always a family member to chime in."

"Well, they're misinformed. Tell your boss that Pinnacle will approve your plans faster than any other company."

Bryce drove back to Suite Seven and reported to Richard what Pinnacle's manager said.

Richard said, "You've found the commercial ethics committee that we want."

Bryce phoned Ted Domain and told him to prepare to do his first review of a clinical trial protocol for Pharm Tech.

Chapter 17: Sebrena's Health

> Pharmaceutical companies want clinical trials to produce results that will help their new drugs succeed in the marketplace. Their aim is not a scientific one, but rather a marketing one.
>
> --Robert Whitaker
> *Our RCT Fetish: How the "Gold Standard" for Research has Led to a Societal Delusion about The Merits of Antidepressantss.*
> Mad in America Blog, September 2022.

Cindy McMullen was in charge of testing Fremontria against Haloperidol and a placebo. She was at Suite Seven when she received a phone call.

"This is Jerry Alvadore with Hancox Pharmaceuticals. I need to see your medical records for Sebrena Southwick. Her sister, Margaret, is worried about Sebrena's health."

"I know who you're referring to. We checked Sebrena's medical history, gave her a physical exam, and interviewed her. We did everything we're supposed to."

"Margaret said Sebrena should be treated as a patient, not a test subject. If she's sick, Hancox needs to know. My boss told me to examine her records."

"Can I check them out and tell you what I find?"

"No. My boss told me to get copies of her records. Send me her intake documents."

"We're really busy. Can I send them tomorrow?"

"Okay."

Cindy was looking through Sebrena's medical records when Sandra walked in.

Cindy said, "How could we have missed this?"

Sandra looked at the records that Cindy pointed to.

Sandra said, "A doctor's report says Sebrena has liver cancer."

"Hancox Pharmceuticals wants her intake file."

"Bryce has to approve anything you send out."

Cindy said, "He'll blow up."

"I'll go with you to see him."

"It might upset him that I told you."

"I'll wait outside his door. If I hear anything upsetting, I'll come in."

Cindy took Sebrena's file to Bryce. She left his office door open.

"What's up?" Bryce said.

"Hancox wants a subject's file. I put them off until tomorrow to give you time to look it over."

"What's wrong?"

"A doctor's report says Sebrena Southwick has liver cancer."

"And Pharm Tech accepted her anyway? Is she being tested?"

"We accepted her, and she's being tested."

Bryce got up and closed the door.

"How could we have missed that?"

"I guess our intake team got reckless. Do you want me to remove her from the trial?"

"No."

Bryce thumbed through the file. His hands trembled.

"Liver cancer is mentioned on two pages."

"Should we fax the file to Hancox and let them tell us whether to remove Sebrena from the trial?"

"That would call our quality control into question. We have a reputation to protect."

Bryce ripped out the pages that referred to liver cancer. He handed the file back to Cindy.

"Fax it to your contact person. We're making it easy for Hancox."

"What happens if Sebrena dies? It could come to light that we knew about the cancer."

"Leave that to Dr. Martinson and me. We need to protect our trial. You can help by keeping quiet."

The next morning, Cindy faxed Sebrena's intake records to Alvadore.

***

On the fourth day of the trial, Cindy found Sebrena unresponsive in her bed.

Cindy called Ira on the intercom.

“Sebrena passed away. A pathologist will record the time of death. When that’s done, I’ll need help in preparing her for the morgue.”

“I’ll be right there.”

A pathologist analyzed the death and recorded vital information. Cindy and Ira delivered Sebrena’s body to the morgue. Cindy went to Bryce’s office in Suite Seven.

“Have you heard about Sebrena?”

“What happened?”

“She died. So, if she was taking Fremontria, we have a complicated case. We won’t know if the drug, the cancer, or a mix of the two killed her.”

“People die whether they’re in drug trials or not. What happens now depends on Dr. Martinson. Work with your other subjects.”

“Are you saying I’m off the case?”

“That’s what I’m saying. The subject has died, so a different set of experts takes over: a pathologist, a mortician, and a priest.”

Bryce phoned Richard. They agreed to meet at the hospital to assess Sebrena’s medical records.

Later, at the hospital, Richard said, "Sebrena was taking Fremontria. We won't report this to the FDA. I want you to fix the data so they show that Sebrena was in the placebo group. That's the easiest way to fix this problem."

"What about her sister? Margaret might say we didn't listen to her warning."

"I'll take care of any concerns she might have."

***

Richard went to Marget's home.

"Hello. I'm Dr. Richard Martinson, Chief of the Psychiatric Department at Green State Medical College. We were caring for your sister."

"You've come in person. This means Sebrena has died."

Margret turned around and led the way into the house. She motioned for Richard to follow.

"We took to heart your notice that Sebrena had liver cancer. I presume you didn't see her at the hospital."

"Your staff discouraged visitors, saying they can mess up a scientific study. Also, Sebrena was trying to keep her cancer a secret. She seemed to think the disease would go away if she denied its presence."

"As soon as we learned she was ill, we kept her from entering the trial, and the cancer doctors treated her as a patient."

"Was she in the hospital as a patient when she died?"

"Yes. The hospital gave her state-of-the-art care. I hope you get comfort from that fact."

"It helps. I didn't want her taking experimental drugs."

"MY goodness, no, we'd never do that. Contact me personally if you want more information. I'll call you tomorrow to see if you have questions."

"I'm glad that Sebrena had such good care."

Chapter 18: Saying Good-bye to Hiram

> I could well imagine that I might have lived in former centuries and there encountered questions I was not yet able to answer; that I had been born again because I had not fulfilled the task given me.
>
> --Carl Jung
> 1875-1961
> Founder of analytical psychology

Hiram died while harvesting ryegrass. His wife phoned their son.

"Bryce, I'm at Mercy Hospital. Your father passed away."

"Oh, no. How did it happen?"

"He was driving a combine and had a heart attack. Mr. Singleton found him."

"I'll be home by evening."

Bryce went to Richard's Suite Seven office.

"I need time off to help Mom make funeral arrangements. I might move her in with a neighbor or into an apartment in town. She shouldn't live by herself."

Richard said, "When I fell, your dad helped get me to a hospital. It was raining, but he rode in the pickup bed to give me room in the cab. He sat in wet clothes at the hospital while doctors treated me. I'll always remember his kindness."

"Can I take a week off?"

"Take as much time as you need."

"What about the trials?"

"Loretta will manage them."

***

The funeral was in an old wooden church in Shedd. After the service, the mourners went to a cemetery where there was a graveside service.

As the service ended, mourners stepped back from a tall man. Richard pulled back his overcoat, drew a sword, and swooped it through the air.

"Hiram rescued me when I fell in battle. I'll protect him as he enters the hereafter."

Ethan went to his son, reached up, and grabbed the hand that held the sword.

"You're Richard Martinson. You're at Halsey, where you grew up. It's 1990."

"Get your hands off me."

"Who are you?"

"Richard the Lionheart."

"You're Richard Martinson, my son. Give me the sword."

Richard relaxed his grip. Ethan took the weapon. He led Richard to the family pickup. Flora waved goodbye as the family left for home.

The mourners began to babble. Bryce raised his arms and shouted.

"You all know Ethan and Flora. Most of you remember Richard, their son. When he hurt his head in a fall, he might have died if it were not for Hiram's help."

The mourners went to Woody's Bar & Grill in Halsey. Woody had reserved a dining room for them. He came from the kitchen and sat with Sarah and Bryce.

Chapter 19: Akathisia

> After taking Paxil 10 mg for 3 days, Mr. Lacuzong drowned his two children and himself in a bathtub. The case was eventually "resolved" without the details of the settlement being made public. The drug company denied all allegations.
>
> --Peter Breggin
> *How GlaxoSmithKline Suppressed Data on Paxil-induced Akathisia: Interpretations for Suicidality and Violence*, 2006

Hancox Pharmaceuticals wanted its new drug, Fremontria, tested against Imipramine, a tricyclic antidepressant that had been on the market since 1957. Since it was no longer protected by a patent, it was unlikely that anyone would defend Imipramine.

Hancox was investigating Pharm Tech's qualifications for doing the trial. Phillip Kragen, the company's vice president for marketing, had come to Suite Seven to meet with Richard and Bryce. With him was technical writer Wayne Endicott. Kragen had the floor.

"I want assurance that you're not secretly recording this meeting."

"We're not," Richard said.

"I want assurance that, except for communicating with Wayne or me, you'll never report anything said here."

"You have my word."

Kragen looked at Bryce.

"I want to hear it from you, too."

"I'll keep our discussions secret."

"Can you keep your staff from betraying our confidence?"

"Absolutely," Richard said. "Bryce and I tailored our procedures so that none of our staff understands the full trial. Tasks are specialized and repetitive. Each staff member sees only the step or two that he takes himself. Each has a small part of the action, which he does repeatedly, like drawing blood, taking blood pressure, or handing out medication.

"Bryce and I are the only ones who see all the data. Staff do statistical tests of isolated parts of a study but never the whole operation."

Kragen said, "If we hire you, it'll be on the condition that Wayne works with you analyzing and reporting the trial. Is that acceptable?"

"Absolutely," Richard said.

"Here's what I want," Kragen said. "Wayne and you will make sure the data are the way we want. The two of you will create a master table and put the figures into whatever column and row makes Fremontria look the best. When we agree on the content of the table, we'll automatically agree on what our technical paper will say."

Hancox hired Pharm Tech. The trial at the VA Hospital ran for six weeks. One hundred test subjects had been diagnosed as depressed and took Fremontria. A separate one hundred depressed subjects took Imipramine. Ten subjects took a placebo.

Five healthy volunteers took Fremontria. Five other volunteers took Imipramine, and a final five took a placebo.

Ten of the subjects who took Fremontria tried to kill themselves. Two succeeded. One of the two suicides had been among the five healthy volunteers who took Fremontria. This test subject had never been depressed.

When the trial results were in, Endicott came to Suite Seven. He and Richard were in Richard's office. A computer screen showed two identical tables of data on a split screen. The two men didn't change the table on the left side of the screen. It showed the data as they had been produced in the trial. This table would remain unchanged, so the two men would have an exact record of reality.

Richard and Wayne changed the table on the right side of the screen to make Fremontria look safer and more effective than it really was. The revised table was to appear in any technical paper that Pharm Tech might publish. Any piece of data might be changed more than once as the men revised their thinking. Both men typed in changes as they revised the outcome of the drug trial.

Wayne said, "Let's move three of the suicides from the Fremontria column to the placebo column to make it appear that our drug is safe and that the risk of suicide in placebo is high."

Richard typed a "3" into the placebo column and in the row marked "suicide." Then Richard reduced by three the number of suicides that occurred in the group who took Fremontria.

Wayne said, "What are we going to do with the healthy volunteer who killed himself? Since the man had been healthy, we can't claim that depression killed him."

"Let's relocate the dead man from the healthy volunteer group to the Imipramine group. The double-blind requirement means that nobody knew which test subject was taking which drug. We know now because we broke the blind, but we haven't told the FDA or anyone else. So, we can report that our suicide fatality was taking Imipramine.

"The dead man was not politically connected. He was homeless, so no one cared about him. The FDA has no way to figure out what happened."

Wayne entered the suicide into the Imipramine group.

Wayne said, "It doesn't matter that he once was a healthy volunteer. He's dead now, and we can't let him screw up our marketing of Fremontria."

Richard asked, "Is there anything else you'd like to change?"

"We can't admit that Fremontria causes akathisia. Many healthcare professionals know that akathisia is linked to suicide and homicide."

"Of the one hundred subjects who took Fremontria, thirty contracted akathisia. How do we handle these individuals without admitting that Fremontria caused akathisia?"

Wayne said, "Hancox has a rule that we don't discuss in the narrative any condition that occurs in fewer than 5 percent of the population that received treatment. We only list such conditions in our report's table. We do no more than that."

Bryce said, "When we include the healthy volunteers, 105 subjects took Fremontria. Thirty got akathisia. Thirty is more than

28 percent of 105. That's way over 5 percent. So Hancox's rule doesn't help us."

Wayne said, "Akathisia can be called jitteriness, irritability, inner restlessness, dysphoria, psychomotor unease, and agitation. That's six other names for akathisia. Divide 28 by 6, and you get 4.7 percent. Use these other names for akathisia and divide the cases among them.

"Get the number down to under five percent in each category we invent. Note them only in our table. We'll break the akathisia-suicide issue into separate, smaller, renamed categories and make each seem insignificant."

Richard scooted his chair around to make it easier to enter data into the table. He followed Wayne's instructions.

Wayne said, "Fremontria doesn't look so lethal now."

"What if a test subject's family member sues us? Can the plaintiffs' investigators figure out what we've done?"

"That's a risk we'll just have to take."

Richard saved their work on a flash drive and handed it to Wayne. They agreed on a prospective publisher.

Chapter 20: Keep the Drug Company Happy

> Trinkets bloom into meals at fine restaurants, meals grow into speaking fees, speaking fees morph into ongoing consultations and member-ships on drug company advisory boards—positions that command up to six figuers a year.
>
> --Jerome P. Kassirer
> *On the Take: How Medicine;s Complicity with Big Business can Endanger Your Health.* Oxford 2005

Bryce was running drug trials for the Hancox Company. His assignment was to bring Richard results that indicated Fremontria was superior to Stelazine and to a placebo.

If Bryce ran enough trials, it was likely that a few would favor Fremontria. Richard could submit these trials for publication and ignore the ones that had gone badly for the drug.

Coordinator Betty Schroder was assisting Bryce in managing the drug trial named Fremontria One. They were in Bryce's office in Suite Seven.

"This is your first time to assist me in managing a drug trial."

"Do you have any advice for me?"

"Keep the drug company happy. That's who pays us. They'll pay up as long as we make their drug look good."

"What's our relationship with the drug company as compared to our relationship with the FDA and with the test subjects?"

"With the FDA, we're like an attorney arguing a case in court. With the test subjects, we act more like a doctor. But the key to running successful trials is to help drug companies get the results they want. Just remember the old saying, 'He who pays the piper calls the tune.'"

"What other rules are important?"

"The FDA's most important rule is that for approval to sell their product, drug companies need to submit only two trials wherein the experimental drug is more effective and safer than a placebo. We don't have to show that our drug is better than one on the market. And we can run as many trials as we want in getting two that are better than the results produced by a placebo.

"Let's say that in thirty of our tests, our drug is a flop. But in two trials, our drug outshines the placebo. The FDA rules say we can submit the two winners and forget the thirty losers.

"The FDA seems to say, 'Run statistical tests that are based on the idea that you're only doing one trial. If you get good results, publish the trial. If you get bad results, do another trial and run the same statistical test as if you were doing only one trial. Do this until you get a second trial that shows favorable results. Publish your two successful trials."

"Really? I can ignore the trials that show negative results for the company's drug?"

"That's right. The FDA almost never asks for results from negative trials."

***

Betty ran a six-week clinical drug trial that compared Fremontria to Stelazine and to a placebo. Fremontria was less effective than Stelazine or the placebo.

The safety results for Fremontria were worse than those for effectiveness. Of the 100 test subjects who took Fremontria, seven attempted suicide, and two died. There were no suicide attempts in the Stelazine test group. In the healthy volunteer group that was taking Fremontria, one person died of suicide.

Betty phoned the Hancox Company and spoke to case manager Larry Ralston. She said Fremontria had done poorly in safety and effectiveness.

Larry said, "Run another trial, Fremontria Two. Lower the dose of Fremontria by a third to see if that lowers the number of suicides."

Pharm Tech enrolled enough test subjects to run a new head-to-head contest between Fremontria and Stelazine. Even with the lower dose of Fremontria, the outcome was similar to the first trial. Again, Betty talked with Larry Ralston.

A third trial also produced dismal results. Fremontria was strongly associated with suicidality.

After the fourth trial, Larry said it was time to do something different.

"It's time we did something more. Move the suicides to different columns in our table. Move suicides from the Fremontria column to the Stelazine column. And put a couple of suicides in the placebo column to make the natural risk of suicides appear high.

"We need to beef up our objective of making Stelazine seem riskier and Fremontria less so. Fax me a copy of your proposed table."

Betty manipulated the figures in the table. She faxed it to Larry. He approved it.

Betty gave Bryce the original paperwork and the data set she had modified. He looked through the papers and told Betty to fax them to Richard's home. Four days later, Richard phoned Betty.

"I read the papers that you sent me. Then, I wrote a draft paper. I put the best face on the suicides that occurred. My draft paper makes Fremontria look good."

Richard faxed his draft paper to Betty. She showed it to Susan Goodwin.

Susan said, "In the last trial, you showed me that Fremontria had caused eight suicide attempts. You moved five to the Stelazine column and three to the placebo column. Richard's draft paper reflects those changes. He's your boss. You never would have cooked up this approach if Bryce and Richard had not pressured you."

Betty said, "Our clinic is helping the drug manufacturer deceive the public. I never should have gone along with my bosses. I thought it was the best way to advance my career."

Chapter 21: Bryce worries even more

Genius sometimes consists of knowing when to stop.

--Charles DeGaulle
1890-1970
French military officer and statesman

Bryce was in Suite Seven, meeting with his staff. Sandra was reporting on the status of test subjects. She was describing how well each was doing after taking pills for ten days.

"Test subject number four is a male, twenty-five years old. The medication seems to have no effect on him. He visits with whoever has time for him. Subject number five is a different story, and I'm worried about her. She—"

Bryce cut in.

"Dr. Martinson and I want to know how many subjects you enrolled in trials last week. And if you found new ways of getting people to sign up. Don't waste our time by telling us the details of how Mr. Smith or Mrs. Jones is doing."

Betty said, "But we have to care about our test subjects."

"Yes, but if we don't recruit people, we'll lose profit, and profit is what enables us to operate. Let's not use meeting time to discuss the details of how a certain person is doing."

The meeting ended. Bryce went to Richard's office.

Bryce said, "With respect, how long can we keep this up?"

"Keep what up?"

"We're deceiving the medical college, drug companies, test subjects, and regulators. We're doing drug trials on the cheap while charging the drug companies the highest prices possible. We're falsifying most trials from design to analysis. We can't do these things forever. We'll get caught. I'm worried about going to prison."

"I've considered this. We should estimate how long we can operate this way. We should bail out before we get caught, taking our money with us. I think we can operate this way for three years."

"It's good to hear that you have a plan. I can agree to three years as long as we coordinate our departures."

"We'll be in touch with each other at all times, just as we've done since we were kids."

***

Betty Schroder and Cindy McMullen were in Suite Seven.

Cindy said, "Pharm Tech pays its test subjects so well that I'm entering a trial myself."

To qualify for trials, the Pharm Tech staff recorded their own medical data, including blood pressure levels, blood sugar levels, and electrocardiograms. They traded samples of urine, blood, and cerebrospinal fluid among themselves so they could appear to qualify for trials of drugs. This tactic also helped them enroll friends and families.

Phar Tech members often participated as test subjects in clinical drug trials. They would provide evidence of illness to regulatory agencies. Any staff member with a medical condition could receive payment from Brice for samples of blood, urine, or cerebrospinal fluid. This tactic helped Pharm Tech enroll families and friends in clinical drug trials.

Richard acted unaware of what his staff was doing. He stayed above the fray. Bryce couldn't use the same tactic because the entire staff saw him perform his duties every day. They knew he was involved with everything that went on in the clinic.

Susan told her co-workers, "Talk about willful ignorance. Richard escapes accountability while Bryce is considered guilty as hell."

Chapter 22: Help was too late for Cindy

> Patients suffering from akathisia often use electrical metaphors such as "electricity going through my veins" or "shocks in my head." Words like "excruciating torture" and "indescribable" are commonly used. Patients often say they would rather die than live with akathisia....
>
> These individuals seem to be describing physical phenomena as if they are being tortured from the inside out.
>
> Peter Breggin, MD
> *Brain Disabling Treatments in Psychiatry*, 2008

Cindy McMullen had been a student at Northwest Christian College in Eugene. She left college temporarily to earn money so she wouldn't have to work when she returned.

Betty Schroder was Cindy's best friend. She was the coordinator of an upcoming test of Fremontria against Quetiapine. Hancox Pharmaceuticals was trying to get the antidepressant approved as an antipsychotic. Betty and Cindy were in the drug-testing ward of the VA Hospital. Betty was entering Cindy into the trial.

Cindy said, "I'm unethical. I'm working for Pharm Tech, but at the same time, I'll serve in a clinical drug trial."

"Don't worry. You're volunteering, and that's all that matters. You can get your regular salary while you get paid for serving in the drug trial. Bryce knows what you're doing, and he's okay with it."

"I don't feel right about it."

"It won't help to fret about it. You need the money so you can return to Bible school. You'll help people after you graduate.

You're doing this for a good purpose. We don't have many ways to get ahead. "

Cindy stayed in the hospital at night and worked her regular shift during the day.

The second night that Cindy was in the hospital, a nurse on the VA staff approached her.

Vera said, "Cindy, how can you work for Pharm Tech and serve in a drug trial at the same time?"

"I'm meeting requirements."

"Since you're in the drug trial, aren't you supposed to hang out at the hospital instead of working a shift?"

"Why do you care?"

"I want to enter a trial. What's the best way for me to get accepted? Can I do the same thing that you are?"

"I'm a coordinator of some of the trials. I'll show you a list of upcoming trials and we'll pick one for you to serve in."

***

The test protocol included two weeks in which the test subjects took no drugs. This was to clear drugs out of the test subjects' bodies to avoid confounding the trial results. Then, the subjects would take the experimental drug, the comparator drug, or the placebo. It happened that Cindy had been given Fremontria.

When Cindy stopped taking Fremontria, it quickly left her body, causing withdrawal symptoms. She got up and paced in the hallways. She was too sick to associate her pain with the drug she had taken. She used the bedside pager to call for help.

Vera appeared.

"I can't stop moving. I want to end it all," Cindy said.

"Sounds like akathisia. I'll get propanol. Get back in bed."

'Hurry."

Cindy got back in bed. Vera left. Cindy got up.

She ran to the windows of the waiting room. She picked up a chair and threw it against a window. The chair bounced off the glass, but cracks appeared. Cindy threw another chair, cracking the window some more.

She used a third chair as a battering ram. She charged the window, crashed through it, and fell to the parking lot.

Vera returned, carrying a syringe. A cold wind rushed through the ward. She looked out the broken window at Cindy's body on the sidewalk below.

Chapter 23: Richard gave speeches

A theory that is wrong is considered preferable to admitting our ignorance.

--Elliot Valenstein
Blaming the Brain: The Truth about Drugs and Mental Illness, 2002

Alphatron's vice president of marketing, Clarence Nysather, was speaking to Richard in Suite Seven.

"I'm offering you the position of key opinion leader for my company."

"Pharmaceutical companies usually select a department head from a big university like Harvard."

"Don't sell yourself short. You've established a national reputation as an expert on psychotropic medications. People know your name. In marketing, that counts."

"I accept your offer. Tell me what you expect so I can deliver for you."

"You'll persuade healthcare professionals to prescribe Alphatron drugs. You might give a talk to a gathering assembled by Alphatron. Or you might present at a convention sponsored by another pharmaceutical company.

"You don't want to come across as a salesman. Your talk must be entirely about a mental disorder. You should not mention a drug. The question-and-answer session is when drugs will come in. Someone in the audience will ask how to treat the disorder. We'll have an Alphatron employee planted in the audience to ask questions. You'll have to answer the question, right? You'll lead without appearing to."

"Will I plant the person in the audience?"

"No. A drug rep will do that. If you rehearse questions and answers, it could sound contrived. You'll sell the drug by selling the disorder. Persuade the audience that the disorder is terrible. You can present rates of death of the disorder that you're describing.

"Maybe you'll present a study of some problem that doctors have overlooked. You'll show that the disorder is so severe that healthcare professionals must do everything they can to recognize, diagnose, and treat it. When asked, you'll name a medicine which is an Alphatron product."

***

Richard spoke at conventions. Alphatron measured how persuasive he was by monitoring the sales of prescriptions. The company got prescribers' names from the convention's registration sheets and purchased prescribing records from a data broker licensed to buy and sell prescriptions. In those records, it found the convention-goers' names and the drugs they had prescribed. By tracking prescriptions before and after a convention, Alphatron could see that Richard's presentation had greatly increased the sale of drugs.

Alphatron sponsored a medical drug convention in San Diego. Richard gave a talk entitled "The Chemical Imbalance Theory." Of the three concurrent sessions, Richard's had by far the biggest audience.

Richard stepped up to the lectern.

"Two hypotheses make up the chemical imbalance theory. These hypotheses concern chemicals called neurotransmitters that pass messages between nerve cells. One hypothesis is that psychotic disorders are caused by too much dopamine, a neurotransmitter. The other hypothesis is that too little of the serotonin neurotransmitter causes depressive illnesses.

"So, a patient can have too much or too little of a neurotransmitter. That's the essence of the theory that you must bear in mind when you diagnose and treat your patients."

Questions came from the floor.

"Wait, let me finish my talk; I'll answer all your questions. Antipsychiatry activists argue that the chemical imbalance theory is wrong, but the theory is sound science as borne out by clinical drug trials."

A man spoke from the audience.

"What is the correct level of dopamine or serotonin? What level should we try to establish?"

"When we lower the dopamine level, the psychotic patient gets better," Richard said. "When we raise the serotonin level, the depressed patient gets better. Experimentation may be necessary, but you'll get the right dosage."

"Are you saying psychiatry determines the dose by trial and error?"

"You're putting words in my mouth."

"I'm trying to figure out how you guys operate. Before drug companies and psychiatry fell in love with clinical drug trials, doctors used a concept called 'challenge-dechallenge-rechallenge.' They used clinical judgment when interpreting the results. Now psychiatrists rely solely on clinical drug trials."

"There is no better way," Richard said. "Researchers have developed rating scales for mental illnesses so clinicians can quantify their analyses."

The questioner in the audience stood up.

"Psychiatrists imply that a six-week trial will predict benefits for the patient years later. That's a lie, and you know it."

"Are you going to let me finish my talk?"

The man sat down. Richard resumed his presentation. The man stood up again.

"Long-term outcomes are the best measure of whether psychiatry is helping us. So-called modern psychotropic drugs have been sold in the US since 1955, when Thorazine came on the market. If the new drugs were effective, the prevalence of mental disorders would have declined. Instead, disorders increase in number every year.

"The so-called medical drugs are causing disorders to worsen. Drugs are turning episodic disorders into chronic ones. A prime example is depression. Drugs are the problem. Psychiatrists prescribe them because the American Psychiatric Association tells them to. Selling drugs is highly profitable. Folks, it's all about money."

A groan rippled through the audience.

Two bouncers grabbed the man's coat at the shoulders. He tried to stay in place. A third bouncer hit him on the head with a billy club. He slumped over. The first two bouncers held onto his shoulders while the third one grabbed him by the ankles. They carried him through the crowd to the street and laid him on the sidewalk.

A theater manager said, “Don’t let anyone rob him.”

After a few minutes, the man sat up. The bouncers went back into the convention hall.

After his talk, Richard found Clarence Nysather looking at displays in the lobby.

Clarence said, “My men followed the activist’s cab. We’ll find out who he is, and we’ll investigate him.”

“Have you run into him before?”

“I don’t know him.”

“Mr. Nysather, I’m resigning.”

“Don’t. Things got a little rough. If you stay on, I’ll raise your pay.”

“How much?”

“I’ve been paying you $1,000 per talk. I’ll make it $1,200.”

“Make it $1,500.”

“Okay, I’ll do it. You ran into a problem, but like a good opinion leader, you’re making the incident work in your favor.”

Chapter 24: Neva signed herself out

> Over the past two decades, the pharmaceutical industry has gained unprecedented control over the evaluation of products. Drug companies now finance most clinical research on prescription drugs, and there is mounting evidence that they often skew their research to make their drugs look better and safer.
>
> --Marcia Angell
> *Industry-controlled Clinical Research: A Broken System.* JAMA, 2008

Neva Robards came to Suite Seven. She was answering a classified ad for healthy volunteers in a Pharm Tech trial of Gabicon. The FDA-approved anticonvulsant was to be tested for treating psychosis.

Loretta Getz interviewed Neva.

"You're answering our ad in *The River Town Times*."

"I want to participate as a healthy volunteer. I need money for college tuition."

"If we accept you, you'll receive $150 a day plus meals and lodging. You might receive the test drug, the comparator drug, or a placebo. Or you might receive one kind of pill and then another kind."

"From what I read about clinical trials, I understand that I won't know which pill I'm taking. Even the doctors won't know until after the trial."

"It's called double blinding. The study design also bars anyone with an illness from serving as a healthy volunteer. Your intake application says you've never had psychosis."

"That's right."

"We don't want to find out later that some doctor once diagnosed you as having a mental disorder. Is there anything like that in your background?"

"No."

"Do you have questions?"

"I hope the scientific knowledge that will be gained will help others."

"Your participation will help others. Testing might even help us discover a wonder drug."

"Will I be restricted to the hospital?"

"No. As a healthy volunteer, you may come and go by signing yourself in and out. The drug-testing ward is on the twelfth floor."

Loretta submitted Neva's application to Bryce. He enrolled her in the drug trial.

Twelve days later, the drug test had gone badly for Neva. It was a Tuesday evening. She paced the hospital halls. She went to the nurse's station but didn't find anyone to help her. She went downstairs. Her movements gave her a little relief. She signed out and ran into the rain.

Neva jogged to the nearest highway overpass. A fence along each side of the bridge prevented her from jumping to the highway below.

She jogged two miles to the Morgan Street Bridge over the Willamette. Fences lined both sides of the bridge. She found a hole in the southside fence and squirmed through it. Jagged wire ripped her jacket and gouged her back.

She was outside the fence, above the river channel. She clutched the wire mesh and inched along the edge of the bridge, moving toward the west bank. She worked her way along the bridge until she was above the shoreline. She released the wire and fell to the rocks below.

***

On Wednesday morning, Sandra took medicine to Neva's room. Neva was not there. Sandra delivered medicine to test subjects in other rooms and then returned to Neva's room. She was still missing.

Sandra went to the nurse's station.

"Have you seen Neva Robards?"

A nurse said, "Check the cafeteria."

Sandra went to the cafeteria but did not find Neva. She went to Bryce's office.

"No one has reported her missing. Check the ledger at the lobby. Maybe she left the hospital."

Sandra checked the sign-out register. Neva had signed out at 9:00 the night before. Sandra phoned Neva's mother, Rebecca.

"Mrs. Robards, I've been unable to find Neva. Is she at home?"

"No."

"Do you have an idea where she might be?"

"No. This is not like her."

"Don't worry. She'll show up. I'm just checking."

Two hours later, Rebecca called Sandra.

"Did Neva show up?"

"No. A hospital register says she left the hospital at nine last night."

"I'm calling the police."

At the River Town Police Department, Lieutenant John Hanlon took the call.

"The Department doesn't consider anyone missing until they've been gone for 48 hours."

"Neva is in a clinical drug trial at the VA Hospital. A drug might have disturbed her state of mind. Waiting doesn't make sense."

"I'll tell my captain. Maybe we can start to search sooner than usual."

***

The next morning, two uniformed police officers came to Rebecca's house.

Lieutenant John Hanlon said, "Police found the body of a young woman below the Morgan Street Bridge. She matches the description of your daughter that you gave us."

Rebecca crumbled into an armchair.

Sergeant Anna Carpenter said, "Mrs. Robards, are you all right?"

Rebecca placed her hands over her face. She rocked back and forth.

Hanlon said, "Can you go to the morgue to see if the deceased is your daughter? We'll take you, or you can go on your own."

"I'll go on my own."

An hour later, Rebecca arrived at the morgue. Lieutenant Hanlon and a mortuary assistant met her at the front door. They led her to a huge, cold room. Mortuary cabinets lined two walls.

The assistant opened a cabinet and pulled out a gurney. He drew back a sheet that covered a body. Rebecca slumped to the floor. The assistant used smelling salts to revive her. Hanlon helped her to a chair.

***

Officers Hanlon and Carpenter went to Suite Seven. Bryce invited them into his office.

Hanlon said, "We'd like to talk to Dr. Richard Martinson."

"He's not here today. I'm in charge when he's gone."

Hanlon said, "Police found the body of a woman yesterday below the Morgan Street Bridge. She was identified as Neva Robards. Was she in one of your drug tests?"

"She was a healthy volunteer in a clinical drug trial."

"Could a drug have caused her to commit suicide?"

"We protect the participants in drug trials."

"What will you do now?" Hanlon asked.

"We'll contact the drug maker and the FDA."

"Here's my business card. Call me if you see anything suspicious."

The police left.

Bryce phoned Richard and described the police visit.

Bryce said, "We need to notify the authorities."

"I'll make the calls. I'll explain the drug test protocol. You can listen in."

Richard arrived. He opened the double-blind records.

"Neva was in the healthy group that took Gabicon for ten days. Then, she entered a washout period, after which she was scheduled to take a placebo. She was on the second day of the washout period when she jumped from the bridge."

Bryce said, "Sounds like withdrawal symptoms."

"Find reasons why Gabicon couldn't be the reason. Rebecca might sue for wrongful death, and the FDA will ask questions. We need to keep Alphatron as a client."

"Gabicon has a reputation for causing withdrawal symptoms. It leaves the body quickly when the individual stops consuming it."

Richard said, "I want you to dispute all such claims."

***

Nelson Hubble, an FDA compliance officer from Salem, interviewed Richard and Bryce.

Richard said, "I'm concerned that Neva's suicide will damage Pharm Tech's reputation."

Nelson said, "The FDA won't even consider that your clinic might be at fault."

"How can you be sure?"

"Our general counsel issued a legal opinion that says clinical drug trial data are commercial information that needs protection. Otherwise, the drug maker in trying to develop a new drug would suffer a competitive disadvantage. The company would have to expend time and money defending its actions about what went wrong, while other companies would have no such obligation. The general counsel does not want to punish the drug maker for trying to win approval for a new drug, for being resourceful. That's how the FDA sees it."

Chapter 25: Urban Legend

> The Texas sharpshooter fallacy is when outcomes are analyzed out of context, giving the illusion of causation rather than attributing the outcome to chance.
>
> --Will Kenton
> *Sharpshooter Fallacy Investopedia,* July 2020

Melissa McConnell was a journalism student at the University of Oregon in Eugene. Her 18-year-old sister, Emma, had died of suicide. A psychiatrist, Peter Bello, had diagnosed Emma as depressed. He said that a chemical imbalance in Emma's brain had caused her to take her life. He had prescribed her Fremontria.

Melissa doubted that Bello was correct in blaming a chemical imbalance. She decided to expose the imbalance theory as unscientific. Her first step would be to write a scientific paper.

Loretta Getz was Pharm Tech's intake specialist for an upcoming clinical trial for Fremontria. Loretta interviewed Melissa.

Loretta said, "So you don't have a doctor's diagnosis, but you believe this clinic will diagnose you as depressed."

"That's right. I've been depressed for about two years, in my opinion. I kept putting off doing anything about my condition until I saw your ad in the newspaper."

Loretta and Melissa talked for a while, establishing rapport.

"Do you have any questions before we complete this form?"

"I read that depression is due to a shortage of serotonin in the brain. I've tried to figure out if that's my problem."

"You'll get an answer if we accept you into our study."

"What's the right level of serotonin?"

"Psychiatry is close to figuring that out."

"You don't already know?"

"We're close to finding out."

"If you don't know the level of serotonin that makes for good mental health, how do you figure out the right dose of antidepressant to prescribe?"

"Experience. The clinician watches to see if a dose of antidepressant helps the patient."

"Sounds like trial and error."

"Some adjustment of doses is to be expected."

"How can I tell if I have a chemical imbalance?"

"We don't know exactly what causes you to have too little serotonin. But social and environmental problems play a role. And someone who suffered trauma as a child is prone to be depressed as an adult. Doctors can't solve all of society's problems, so they put their efforts into resolving the biochemical problems they see."

"How can I tell if I have a chemical imbalance?"

"It's hard for you to tell whether you have a chemical imbalance because the imbalance keeps you from thinking clearly. You need to trust a good psychiatrist like we have at Pharm Tech.

"Here are the signs and symptoms of a chemical imbalance. You might get your days and nights turned around, so you sleep in the day and are awake at night. You might sleep too much or not enough. You might overeat, or you might lose your appetite entirely. You'll feel sad and hopeless. You may feel you're in danger all the time. You'll probably abuse drugs. You'll have trouble concentrating. You might go to lengths to avoid people."

"Wait a minute. Those are feelings that anyone might have for a variety of reasons. The person might have the flu or pneumonia, or maybe a divorce or eviction. Problems like the ones you recited don't prove you have a chemical imbalance."

"You need to have faith in a psychiatrist's ability to diagnose your problem."

"I'm looking for a yes or no from a test. Is there a blood test or a urine test? Something that confirms or rejects that I have a chemical imbalance in the brain?"

"We don't have such a test, but psychiatry is discovering things all the time. We're drawing near a definitive diagnostic test. I bet we'll have one within two years. In the meantime, logic helps us make good diagnoses."

"How do you use logic?"

"When a psychiatrist prescribes an antidepressant, the serotonin in the brain increases, and the patient gets better. The improvement in the mental condition coincides with the increase in serotonin."

"You've implied that psychiatry can't measure serotonin. Can it be measured?"

"I should have said that the psychiatrist presumes there is an increase in serotonin. Anyway, logic tells us that the patient got better because the psychiatrist raised the serotonin. Also, antidepressants have been tested in clinical trials like the one you are about to enroll in. And the statistical tests that are done say that the antidepressants work."

"Sounds logical to me."

"If you sign up for the trial, you'll help psychiatrists develop even better diagnostic methods. Let's fill out the intake form."

Melissa filled in the blanks on the form. She asked for a copy of it.

"We don't give out the intake form."

"Why? It's my medical record."

"Dr. Jewel says to not give it to anyone."

"May I see Dr. Jewel? State law says I can have copies of my medical records."

"Okay, I'll give you a copy. Sign here."

"I'll take the form home. I'll come back in a couple of days and will sign up for the drug trial."

"Please enroll now."

"Why?"

"It's easier. Also, it saves you from having to make another trip."

"I don't mind an extra trip."

Melissa got the form and left the clinic. She gathered more information on the chemical imbalance theory. She wrote her scientific paper and dedicated it to Emma McConnell.

Chapter 26: The Texas Sharpshooter

> A Native American elder once described his inner struggles in this manner: Inside of me there are two dogs. One of the dogs is mean and cruel. The other dog is good. The mean dog fights the good dog all the time. When asked which dog wins, he reflected for a moment and replied, "The one I feed the most."
>
> --George Bernard Shaw
> 1856-1950
> Irish playwright and political activist

Mercantile Pharmaceuticals came to Suite Seven. Harold Bush, Vice President of Marketing, arranged for Pharm Tech to test Truplex for the treatment of psychosis.

Bush said, "Mercantile doesn't believe in double-blinding. We like active management. We want you, the contractor, to know which drug every patient is taking. We want you to monitor what's happening and make adjustments. Making changes in dosing will enable you to get good results."

Bryce said, "Richard and I will know which drug each test subject is taking. We'll keep the information from our staff. No one will know we're violating FDA blinding rules."

"We want you to do statistical tests on everything that can be analyzed. If you come up with a significant statistical finding, go back and rewrite the protocol to make it appear that was the result you wanted to test to begin with. But I'm worried that your institutional review board won't let you change the protocol."

Bryce said, "We foresaw this problem, so we hired a flexible IRB. Our model of operation is the Texas sharpshooter."

Harold said, "Tell me about it."

"A Texan wanted to impress his friends with his marksmanship. He fired his rifle into the side of an abandoned barn. He picked out the tightest group of bullet holes and painted a target around it."

"That's my idea of effective management."

"Many research contractors test subjects who don't have the disease being treated. They create symptoms to make it look like the person has the illness. At Pharm Tech, we don't stop there. We keep bodily fluids in refrigerators. We store blood, urine, and cerebrospinal fluid that meet lab values. We use these specimens so that test subjects can persuade authorities that they have certain disorders.

"For example, we pay Helen, a diabetic staff member, $25 for each jar of urine she brings in. We divide the urine among several staff members, and we label each sample with an employee's name. Now, we're prepared for an inspection by a regulatory agency. Most research contractors don't go this far."

***

Sandra was promoted to a more technical position. She was meeting with Bryce at Suite Seven.

She said, "You should publish all the trials, the two with positive results and the twelve with negative results. If you publish only the trials with positive results, prescribers and patients will think Truplex is safer and more effective than it really is. "

"Pharm Tech will publish any trials that favor Truplex and any in which we can spin the results to make Truplex look good. The FDA doesn't require negative results to be published, and neither does any scientific journal that I know of. If we do what you're recommending, we'll lose clients."

"The ethical thing is to publish all trials regardless of their outcome."

"We're in this business for financial gain."

"Findings are distorted when negative results are hidden. Doctors decide on faulty evidence which treatment is best."

"I put our drug makers first," Bryce said. "In fact, I intend to publish the positive results more than once."

" You're going to get a false positive if you give yourself multiple chances but are using statistical tests that assume you had only one opportunity."

Bryce's face flushed.

Sandra said, "If you flip a coin enough times, you'll get four heads in a row. That's not the same as getting all heads in four flips. With regard to our Truplex tests, our report should say that we flipped the coin fourteen times to get it to land twice the way we wanted."

"I'm close to firing you. Shut up."

***

Bryce couldn't legally prescribe drugs. Most people incorrectly assumed he was a physician. But the "doctor" part of his title referred to his PhD in pharmacology. He signed Richard's name on prescriptions and medical records.

Rosemary gathered a stack of documents. She toted them to Bryce's office. He signed them while listening to a baseball game on his radio.

Chapter 27: Reincarnation

> A Native American elder once described his inner struggles in this manner: Inside of me there are two dogs. One of the dogs is mean and cruel. The other dog is good. The mean dog fights the good dog all the time. When asked which dog wins, he reflected for a moment and replied, "The one I feed the most."
>
> —George Bernard Shaw
> Irish playwright and political activist
> 1856-1950

Intermittent sections of the Greenway Trail lined both sides of the Willamette River from Cottage Grove to Portland. The Greenway included boat landings, parks, and footbridges over the river.

Richard and Bryce left Suite Seven at noon and took a walk on the Greenway. After walking a mile, they sat on a park bench.

"Do you believe in reincarnation?" Richard said.

"No. Why do you ask?"

"It's a concept that explains my identity."

"You were raised Christian, more or less. Why would you think a Buddhist concept explains who you are?"

"Buddhism is not the only religion that believes in reincarnation. Others include Hinduism, Jainism, and Sikhism. And some nonreligious people believe they lived before."

Bryce said, "I've known people who said they'd lived before. All said they had been kings or presidents or great philosophers.

None claimed to have been a regular person. Their boastfulness made me doubt their honesty."

"It's natural not to brag about being a common person and to brag if you were royalty. I focused on how accurately the person could describe life in the past and in distant lands. Historians found some of the accounts to be accurate. The person usually had no conventional way to know what he said he experienced."

"Give me an example," Bryce said.

"Edgar Cayce gave thousands of readings while he was asleep. He did this in America in the first half of this century."

"Researchers have pointed out errors in Cayce's predictions."

"But there were other times when his forecasts were correct. He was educated only through the ninth grade, so it's unlikely that he independently learned all that he spoke of."

"You have a point," Bryce said.

"There must be an explanation for the visions I see every day."

"What do you see?"

"It's hard to talk about."

"I understand how it could be."

"But it's also hard not to talk about it, at least to someone. I thought I might be able to talk with you. We've known each other since we were kids."

"I'm willing to talk with you about past lives and visions. But it's different than when we were kids. You were always the leader in a playful way. Now you're the senior partner in our clinic. We have a formal relationship. It'd be hard for me to be your junior partner and your confidant. I'd have to be two people."

"It's strange you put it that way. That's what I wanted to talk to you about, being two people. At the moment, you're talking with Richard, the psychiatrist, but sometimes I'm someone else."

"Who?"

"You probably know. You were there when I fell."

"Richard the Lionheart?"

Richard nodded. He walked to the river and began skipping rocks across the water. Bryce stayed put.

Richard returned to the park bench.

"I don't want to talk anymore."

The two men walked north, the direction the water was flowing. Four mergansers landed on the water and dived for fish. Soon, they flew away.

Richard said, "We'll never see them again. Our relationship was so brief."

The men walked back to Suite Seven.

Chapter 28: We're coordinators, not nurses

The remedy is worse than the disease.

--Francis Bacon
English philosopher
1561-1626

After a clinical trial, eight survivors remained in the VA Hospital due to the drugs they had taken. Richard and Bryce intended to keep them there until they recovered or died.

Rosemary Adams and Carolyn Goodwin were taking care of the patients.

Carolyn said, "Mr. Hayes is doing poorly."

"I'll take a look at him."

Rosemary and Carolyn went to see Gordon Hayes. Although his eyes were open, he didn't stir when they entered his room.

Rosemary tried to get Gordon to eat pudding. He refused. She tried applesauce. He refused.

"I'm sixty-five years old. I was surprised when your clinic accepted me as a test subject."

Carolyn said, "Were you accepted into the drug trial because you were sick?"

"No. I was accepted as a healthy volunteer."

"You weren't depressed?"

"Not when I entered the trial."

"You got sick during the trial?"

"That's right."

"After the trial, were you told whether you took a drug or a placebo?"

"A technician said I took Gabicon. Life isn't worth living since my wife died. Can you help me qualify for assisted suicide?"

Rosemary said, "Oregon has a 'Death with Dignity' law. Our clinic might help you qualify. But maybe you'll change your mind. Let's improve your health so you'll want to live."

"Even when Ruth died, I didn't feel this bad. I never dreamed anything could be as awful as this disease."

Carolyn and Rosemary went to the nurse's station.

"What should we do?" Carolyn asked.

"I don't know. I hate Richard and Bryce for putting us in this predicament. Neither of us has any serious medical training. I led a Girl Scouts organization, and you were a second-grade teacher."

"Do you know why they hired us for these jobs?"

Rosemary said," Because they can pay us one-fourth of what they'd have to pay a nurse. They don't call us nurses. They call us coordinators. "

"So, without medical training, how do we figure out what to do for Gordon? We're expected to take care of him."

"To make matters worse, our bosses take extreme measures to hold down the cost. Gordon and other patients are costing the drug companies money as they stay in the hospital after the drug trials."

Carolyn said, "Let's use the Hamilton Depression Scale to see how severe his symptoms are."

Rosemary found the Hamilton Scale in a file cabinet. She read the checklist of symptoms to Carolyn, along with the scores that could be assigned to each symptom. There were seventeen items to be scored and four more that provided insight into depression.

Carolyn and Rosemary discussed each of the seventeen items to be scored. Checklist item number three instructed the raters to score tendencies for the patient to commit suicide. Rosemary wrote the item on the chalkboard.

Item 3. Suicide

1. Absent
2. Feels life is not worth living
3. Wishes he/she were dead
4. Suicidal ideas or gestures
5. Attempts at suicide

Rosemary said, "Gordon scored a five. He acted like he was taking the medicine, but he stockpiled the pills and overdosed. The Hamilton Scale considers him deeply depressed."

Bryce came to the hospital later that morning. When he entered Gordon's room, Rosemary asked him to step into the hallway so Gordon couldn't hear them talk.

"Gordon wants Pharm Tech to arrange assisted suicide for him. "

"We won't do it. That could hurt our reputation. How many people know of Gordon's grave condition?"

"Carolyn Goodwin and I know. Gordon's wife is deceased. The intake form says he has no other family. He hasn't mentioned any friends in my presence, although he may have some. Our protocol prevented him from having visitors."

"Is he politically connected?"

"Not as far as I know, although I never asked."

"You seem bothered by something," Bryce said.

"Gordon's severely depressed. I think we should give him just enough antidepressant to stop the withdrawal symptoms and then withdraw the drug slowly. That might be one way we can help him."

Bryce said, "We have options. One is electroconvulsive therapy. Another is cognitive behavior therapy. If we choose CBT, we'll keep it quiet because that is not one of the therapies that Pharm Tech promotes. Someone could claim that counseling had to be used to counteract the failure of the chemical methods that we test."

Rosemary said, "I vote for CBT. Let's see if Gordon can recover without taking any more drugs."

Bryce assigned a psychologist to Gordon's case. The psychologist went to the hospital to talk with Gordon about the therapy that would begin soon. She found him dead.

Bryce changed the records to make it appear Gordon had survived the thirty-day post-trial period. That would make Gabicon look safer.

Chapter 29: The Bully

He did not want to be the victim, so he became the bully.

--Tupac Shakur in speaking of himself
American rapper and actor
1971-1995

As a freshman, Ron Iverson was energetic and intelligent. However, he was bullied by older kids at Portland High School. Elson McDonald was two years older and was bigger and stronger. He excelled at sports.

As afternoon classes were about to begin, Ron went to his wall locker for books. Elson had gone to his locker for gym clothes.

Elson said, "Stay away when I'm at my locker. Wait, so we don't run into each other."

"Advice from a bully," Ron said.

Elson pushed Ron, who fell to the floor.

"Stay down if you know what's good for you."

Ron lay on the floor and waited. Elson walked away with his friends. Ron got up and gathered his books.

***

Elson's bullying tactics took different forms, occurring frequently. Elson was a sophomore and would graduate in three years, when Ron could lose track of him. Ron wanted to settle the score before that time. He decided to build up his body and train in hand-to-hand combat. He took classes in weightlifting, boxing, wrestling, and karate.

Ron watched Elson as he excelled on the football field and in the boxing ring. One day during the boxing semi-finals Elson was expected to lose to a boxer from a Seattle high school. But he won. Elson was preparing to leave the ring. Ron stood fifteen feet away.

Elson said, "Do you want to be next?"

"The time will come."

Elson left the ring and headed for the showers.

Ron's self-defense coach saw the exchange between the two students. He gave Ron advice at their next gym class.

"End your grudge against Elson. He can punch like Muhammad Ali. If you fight with him, he might kill you."

"You haven't been humiliated by him like I have. If you want to help, tell me how to outsmart him."

"I don't want to be accused of helping to continue this dispute. I'll just say that the most flexible fighter usually wins. If Elson wants to fistfight, you need to wrestle. Don't let him determine the manner of the fight."

***

A year passed. Feelings worsened. There were insults and name-calling. Elson said Ron was in for a beating.

One night at a lakeside party, the name-calling turned into physical combat. Elson had drank several beers. Ron watched him interacting with his friends. He waited while the big man drank until he staggered when he strutted about.

Ron stepped out of the shadows.

"Hello, Fatso."

"I've got you now."

Elson threw a punch at Ron's chin, but Ron had dived for Elson's ankles to make a wrestling takedown. He drove Elson's feet out from under him. The big man landed face down on Ron's back. Ron stood up, lifting Elson four feet off the ground. Elson fell, landing on his back. He lost consciousness.

Ron tried to choke his enemy but couldn't get his hands around the man's massive neck. Ron kneeled on Elson's chest and pelted him with lefts and rights.

One of Elson's friends said, "We better get out of here before the cops come."

Ron stood up, releasing Elson from his pinned position. Two of Elson's friends helped him into a car.

***

Ron spoke with his wrestling coach.

"I did what you said. Elson wanted to fistfight, so I wrestled. I took him down by driving his feet out from under him. He threw a punch that helped cause him to fall forward onto my back. When

I stood up, he fell off my shoulders. He landed on his back and got the wind knocked out of him."

"You didn't knock the wind out of him. A fall like that temporarily paralyzes the diaphragm. Elson probably couldn't breathe for three or four minutes."

"I did not box. I wrestled. That's all I did, but I looked like Bruce Lee."

"You did more than that. You saw that Elson was drinking, and you waited until he was soused. His physical and mental abilities were diminished before you took him on. Elson told friends that he thought he was going to die."

***

Years later, the wrestling coach told a friend what he had learned by watching Ron and Elson settle their dispute.

"Ron Iverson not only changed Elson McDonald, but he changed himself, too. Elson was a bully who became a decent person after Ron beat him up. Ron had been a fun-loving person, but he turned bad as his hatred of Elson worsened. His victory over Elson turned him into a bully to be feared."

Chapter 30: Head-to-head Comparison

Why Olanzapine Beats Risperidone, Risperidone beats Quetapine, and Quetapine beats Olanzapine: An Exploratory Analysis of Head-to-Head Comparison Studies of Second Generation Antipsychotics.

— Stephan Heres, et al., 2006
Title of a report in The American Journal of Psychiatry

Alphatron Pharmaceutical's vice president of marketing came to Suite Seven to arrange a clinical trial for Gabicon. Clarence Nysather shook hands with Bryce. They had collaborated on an earlier trial. They sat in easy chairs, and Loretta brought coffee.

Clarence said, "Gabicon is doing well as an anticonvulsant. But as an antipsychotic, it is selling poorly. Moliperdone is outselling it. Gabicon should own most of Moliperdone's market. Iverson Company ads aren't as good as ours. But Moliperdone has caught on with prescribers while Gabicon hasn't."

"While we helped you get Gabicon approved as an anticonvulsant, Moliperdone entered the antipsychotic market. Moliperdone's marketing advantage over Gabicon probably is due to that head start."

Clarence said, "Whatever the reason, we need to make doctors and patients realize that we have the better drug. We need to go head-to-head in a trial against Moliperdone."

"We'll put together a trial."

"I don't know if Gabicon can outdo Moliperdone. I don't want to pay for a trial that shows Moliperdone is the better drug. We can't make everything up because that would be recognizable as fraud. We're safer if we do an actual trial that we manipulate in subtle ways. We need to have real patients react to real drugs."

"There are a couple of things we can do." Bryce said, "We can do pretests with Gabicon in one group and Moliperdone in another. Later, in the actual trial, we'll use any subject who did well while taking Gabicon. We'll leave out anyone who did well on Moliperdone."

"Pretests will enable us to get the results we want in the trial. At the same time, they'll look realistic."

"We also will run a pretest on subjects who take a placebo. We'll leave out of the trial anyone who did well on the placebo. This'll make Gabicon look better."

"I like what you're saying. We'll handpick our test subjects," Clarence said. "Is there anything else we can do?"

"We'll find the doses that patients do well on when taking Moliperdone. We'll cut an effective dose in half, but show it as a full dose. That'll make Moliperdone look less effective."

"Sounds good," Clarence said.

***

Pharm Tech did the pretests. It administered Moliperdone at 4 mg/day instead of the 8 mg that Richard showed in the report he wrote. The clinic falsified the records to make the misdeeds look legitimate.

A year after the drug trial ended, Richard published his paper. It promoted Gabicon and disguised the illegal tactics that Pharm Tech had used. Within a week of publication, a letter came from a law firm representing the Ronald Iverson Company. It was a notice of intent to sue Pharm Tech for claiming that Gabicon was superior to Moliperdone. The notice said that the clinical trial was a fraud.

Richard's lawyer, Jeffrey Nash, set up a meeting with the Iverson Company and its lawyers. Richard and Nash traveled to Iverson's office in Seattle. The Alphatron officers arrived separately.

Ron Iverson, the company's CEO, allowed his visitors to see that he carried a handgun under his suit jacket.

Iverson asked Richard, "Why did you do a head-to-head drug trial? The FDA doesn't require such a test. It allows the research contractor to test the proposed drug against a placebo. That way, no one having a market share gets hurt. Neither does the company seeking FDA approval. You intended to make Moliperdone look like a loser."

Richard said, "We did a head-to-head because Alphatron asked us to. They're simply competing in a tough market."

"Did Pharm Tech do pretests to find subjects who reacted badly to Moliperdone?"

"Absolutely not."

Iverson said, "Did you use the dose you claimed, or did you secretly use a smaller dose of Moliperdone so it would do poorly?"

"We did the trial exactly as we reported."

"The Iverson Company wants you to pay $40 million for the damages you caused. Unless you do, we'll sue you and Alphatron. We'll depose your staff. They might disclose that you used pretests and a low dose of Moliperdone. They might remember things that you claim didn't happen."

Jeffrey asked for a break. He met with Richard and the Alphatron team.

Jeffrey said, "Iverson is a thug, a mob boss. I'm not as worried about a lawsuit as I am about him having you killed."

"What's the best way out of this mess?" Richard said.

"The answer depends on how honest your clinical trial was. Did you do anything that could cause you to lose in court or cause Iverson to have you killed? Level with me."

"We twisted things a little."

'A little or a lot?"

"Okay, okay, okay, we twisted things."

"Settle with him."

The meeting resumed.

Richard said, "Here's our counteroffer. We'll do a free trial for you. We'll put Moliperdone up against a placebo."

Iverson said, "I want you to test Moliperdone against Gabicon without using any tricks, pretests, or falsification of doses. You'll cooperate with an expert from my company, who'll observe the trial. And don't forget the $40 million in damages."

"We won't do all that."

Jeffrey said, "We need another break."

Jeffrey and Richard went into the hallway.

"I repeat that Iverson is apt to have you killed. You crossed a mob boss. Settle with him. Otherwise, you might be dead within hours."

Jeffrey and Richard returned to the meeting,

"Acting on legal advice, I agree to your offer. But you must understand that trials can yield varied results."

"Stop crying the blues. I'll give you thirty days to pay the $40 million."

RIchard said, "I won't cross you. I'm agreeing to run a head-to-head trial at no expense to you and to pay you $40 million."

Iverson shifted the position of his handgun so everyone could see that he was carrying.

Chapter 31: You could get killed

> It is scary how many similarities there are between this industry and the mob. The mob makes obscene amounts of money as does this industry. The side effects of organized crime are killings and deaths, and the side effects are the same in this industry. The mob bribes politicians and others, as does this industry.
>
> --Peter Rost
> A former drug company executive
> http://peterrost,blogspot.dk June 2012

Ron Iverson assigned statistician Dennis Klepper to work with Pharm Tech in analyzing a new trial of Gabicon versus Moliperdone.

Dennis and Bryce were in Suite Seven. Bryce could see the outline of a handgun beneath Dennis's sport coat.

Bryce said, "I had nothing to do with that trial between Gabicon and Moliperdone."

"Does Martinson review everything you do?"

"What do you mean?"

"How much are you required to play by Richard's rules? Pharm Tech broke FDA rules to make Gabicon look better than Moliperdone. But you weren't figuring on Iverson coming after you. You didn't know about his violent background. Now you're boxed in. You broke medical and legal rules to make Gabicon appear to be better than Moliperdone.

"Here's what Ron wants. We need to show that Moliperdone is better in every respect. It acts faster and does not cause side effects. We have to show that patients love the drug."

"I understood that Iverson wants an honest test between the two drugs. But now you're telling me that I'm to favor Moliperdone at every turn. It would destroy Pharm Tech's reputation to say that Gabicon is by far the better drug, then to immediately reverse ourselves by saying that Moliperdone is better in every way."

"You're in a bad spot. If you don't cheat for us now, we'll punish you for cheating in the first trial between Gabicon and Moliperdone. Maybe there'll be fatal consequences."

Bryce said, "But the trial must be believable. It must not be seen as a lie. We have to take into account the existence of a published scientific report that extols the virtues of Gabicon. It's likely that no scientific journal will punish what we write now. Think about that."

"This whole publication business is corrupt. Iverson will get dirt on a high-powered editor somewhere. He'll make that editor's journal publish our paper if he has to use the same tactics on the editor that he's using on you.

"Your job is to produce a trial that promotes Moliperdone. If you don't go far enough, we'll make you try again. You've shown you have the ability to deceive. If you don't go far enough, we'll make you try again."

"We have a reputation to protect. If we don't come across as honest, no journal will accept our paper."

"I'm not forgetting the value of your reputation," Dennis said. "You have a lot more to lose than we do."

"We have an agreement with Iverson that he won't sue us."

"I keep telling you that lawsuits are one thing, but we have other ways of getting what we want. I'm about to show you one of those ways."

"I need to convince you that I'm on your side. I'm just trying to make our paper believable."

"Stop yakking and start doing."

***

Bryce went to Suite Seven and reported to Richard.

'Klepper carries a gun like Iverson does. He threatened to kill us."

"I expected this."

"Why?"

"Since our meeting, I've learned things about Iverson. He was investigated for having people beaten up. He was found not guilty of murder. The beatings and the alleged murder involved drugs – either prescription or street drugs. It appears the not-guilty verdict made him more inclined to use threats of violence against his enemies.

"I wish I'd known of Iverson's violent behavior before we did the Gabicon-Moliperdone trial. I would never have done a head-to-head trial. I blame Alphatron for not telling us because they must have known something about his history. Alphatron was reckless."

Bryce said, "What should I do?"

"We can't end up in court. That would do too much damage to our business. If we don't settle this dispute, Iverson will, at the least, get us into serious legal trouble. More likely, he'll have at least one of us killed. He won't be satisfied with anything less than a drug trial that says Moliperdone is far better than Gabicon."

"We might ruin our reputation if we give Iverson the trial he wants."

Richard said, "We can re-establish our reputation. People have short memories. After a few years, our trespasses would be forgotten or forgiven. The term "Overcome by events" applies here. New issues are always coming up. People have new things to worry about. It might come down to a choice between our reputation and our lives. Iverson and Klepper have us boxed in. I want to get away from them."

***

Bryce met with Dennis the next day.

"When are you acting for Iverson, and when for yourself?"

"That's none of your business. Just take seriously what I say."

"I'm convinced this is serious."

Bryce persuaded Dennis to accept four installment payments. He paid the first payment of $10 million with cash.

Bryce and Richard made Moliperdone look good.

Richard told Bryce, “When you deal with a drug company and money is involved, be aware that you could get killed.”

Chapter 32: Not a Place to Retire

Plans are things that change.

Fujio Cho, Born 1937

Was honorary chairman of the Toyota Motor Corporation

Richard and Bryce ate a catered breakfast at Suite Seven.

Richard said, “Do you want to get rich?”

“Of course.”

“Then why not recruit more test subjects?”

“Things are going well. We’re in black ink, and we’ve been in business for only two years. “

“Six months ago, how many drug companies were you doing business for?”

“I can’t say offhand. I didn’t expect to be grilled this morning.”

“We had six sponsors. How many do we have now?”

“Not as many as you want.”

Bryce got up and headed for the door.

“Get back here and sit down.”

Bryce sat down. He glared at Richard.

Richard said, "What is the key to doing lots of trials? What have I always said?"

"Rapid recruitment."

"That's right. So, let's focus on how fast we can get test subjects. Rapid recruitment has become more important as drug companies make 'me too drugs.' It takes only a small trial to reveal if a drug is highly effective. But a 'me too drug' may be barely better than a placebo. And it takes a big trial to show a significant difference between a 'me-too drug' and a placebo. That's a major reason why we are always forced to find more subjects for our drug trials. If the drugs were not just imitations of other drugs, we could get enough test subjects much easier."

Bryce said, "It makes it even harder to get enough test subjects when drug trials occasionally kill someone. Prospective test subjects hear the stories and refuse to participate."

"You're correct to focus on problems, but let's remember they also provide us opportunities. If we specialize in getting subjects quickly, we can outdo the other contractors who let the difficulties of recruiting weigh them down."

"I'm doing everything I can."

"We need to explore how we can recruit test subjects faster."

"It would help if you came into the hospital more often. I wouldn't be so overwhelmed."

"As head of the psychiatry department, I'm publishing articles. Our publications tell the drug companies that we're experts who can serve them well. It takes time to research and write those papers. I can't do that at the hospital or Suite Seven. When I write, I need it to be quiet. I have to be at home, or at a library, or maybe a park. Alphatron hired me as a key opinion leader. That shows that my methods are working."

"Okay, but don't blame me. Drug trials are hard to run."

"Other than using beautiful women to recruit men to serve as test subjects, tell me of a way to recruit quickly," Richard said.

"Here's a way. There are patients at the Oregon State Hospital who are serving an involuntary commitment. You could go to the hospital and cut a deal with the superintendent to take patients off his hands."

Richard said, "I see. We could take any patient who we could use as a test subject. The person would have to agree to obey the rules of clinical drug trials, or he would be involuntarily committed. But we could be seen as using unfair coercion."

"You could defend against charges of coercion if you made it look like the patient brought up the idea and talked you into it. The superintendent should like the idea because he must have dozens of people who are sick and are depleting his budget. You'd help him relieve any overcrowding that the hospital might be having."

"Maybe I could get the patients delivered to the River Town VA Hospital. Do you have anything else?"

"I'm worried that we can't keep fooling the authorities forever."

"We shouldn't think of this clinic as a place to retire. We should take it for all it's worth, then get out before we get caught. We should take calculated risks until it's time to leave."

"Why are you taking time to write journal articles if you're planning to bail out?"

"Scientific articles got me selected as a key opinion leader. That brings in a lot of money. And Alphatron pays in cash immediately after I give a presentation."

"Do I benefit as well as you? It's my running Pharm Tech that makes it possible for you to research, write, and serve as a key opinion leader."

"You're co-author on most of the papers."

"Does Arnold realize that you see the trials as a short-term affair?"

"Arnold can look out for himself. He takes a big cut without doing much."

"Will you leave the country when you quit this business?"

"No. I have a different plan. What have you learned while I've been writing papers and giving speeches?"

Bryce said, "We can't let a prospective subject deliberate about whether to enroll in a trial. Our coordinator must insist that the prospect sign up immediately, that they have no choice but to participate."

"Another way for Pharm Tech to recruit rapidly is for our staff to sign up as test subjects. We'll get more recruits if our staff members claim to have the sicknesses that we test for."

"That's all right with you?"

"Absolutely."

"Good," Bryce said, "because we're already doing that. Unfortunately, that's what killed Cindy McMullen."

"But we didn't get blamed. That's important. The method is sound. The FDA protects any company that tries to develop a new drug."

Bryce said, "You're changing plans. You're saying we shouldn't retire here. I can live with that as long as we leave at the same time. You can't desert me."

"I won't. This plan enables us to maximize profits and then get out before authorities are alerted. I won't leave the area but will hole up where they can't get me."

Chapter 33: Stay of Commitment

> The death prompted reviews by the state mental health ombudsman and the U.S Food and Drug Administration about the conduct of the [University of Minnesota] and [Stephen Olson] who was [Dan] Markinson's only psychiatrist at the time he recruited him into the study. The reviews and the lawsuit probed whether Markinson was coerced into the study by the threat of commitment to a psychiatric hospital....
>
> --Jeremy Olson and Paul Tosto
> *The Death of Subject 13*
> St. Paul Pioneer Press (Minnesota)
> May 18, 2008

Richard and Bryce continued their meeting at Suite Seven.

Richard said, "The more I think about your stay of commitment idea, the more I like it."

"What do you like?"

"First, there'll always be a supply of potential test subjects. The State Hospital will always have mentally ill people who qualify as test subjects for clinical trials.

"Second, we'll have control over the patient. He or she won't be able to back out the way regular citizens can. If the person tries to back out, we can threaten to send them to a locked psychiatric ward. Or we can send them to such a ward for a while and make them more cooperative.

"Third, the State Hospital will offer us a kind of patient that's hard for us to get otherwise."

Bryce said, “You’ll have to explain that last reason to me.”

“AstraZeneca has asked Pharm Tech to do a Comparison of Atypicals in the First Episode of Schizophrenia. It’s very hard to tell if a person is having their first episode of psychosis. So we’re looking for an especially hard-to-identify specimen. The State Hospital is likely to have compiled a history of whether the person has been sick before. That history could tell us when a person has had previous episodes. Having a history will be very helpful.”

***

Richard made an appointment to meet with Edward Ashton, the Superintendent of the Oregon State Hospital. He drove to Salem and went through the security checks to get into the hospital.

“How can I help you?”

“I’d like to discuss the clinical drug trials that my college does. I'd like to propose a business deal.”

“We participate in clinical trials.”

“Good. Pharm Tech, the college’s clinic, has difficulty signing up clinical test subjects. People think the trials are too risky. They’re actually safe, but I have trouble convincing the public of that fact. The Zonal Company has asked us to do a CAFÉ study. Do you know what CAFÉ stands for?

“Comparison of Atypicals for First Episodes. It must be hard to get test subjects for this kind of study.”

"That's why I'm here. You have patients who are locked up in special units because they're intensely ill. Pharm Tech can use those people as clinical test subjects."

"You're talking about people bound by a commitment order. They're ordered by a court to be locked up because a psychiatrist has determined they are a danger to themselves or others."

"I'm willing to testify that participating in a clinical drug trial could help the patient as much as, or more than, standard treatment. I would ask that a stay of commitment be issued."

Ashton said, "The college would be responsible if anything goes wrong. Will you agree to that requirement?"

"We're responsible for everyone in our trials, which are carefully designed and managed. When we propose a protocol, an Internal Review Board must approve it."

Ashton and Martinson explored the benefits and the risks of the college using in its clinical drug trial patients who were, or were likely to be, committed to involuntary commitment. No one was there to speak for the patients.

The two bureaucrats agreed to try the program on a few patients to sort out any difficulties before they committed to a full-fledged version.

***

At the Oregon State Hospital, Alfred Lane was locked up involuntarily. He had been diagnosed as having his first episode of schizophrenia.

Before he became psychotic, Alfred had been easy to get along with. He had obtained a bachelor's degree in English literature from the University of Oregon.

He was facing a civil commitment hearing. The critical issue was whether Alfred was dangerous. His father, Grant, had given the State Police a statement that Alfred had threatened to kill him with a hunting knife that he kept under his pillow. A psychiatrist had talked Alfred into surrendering the knife.

Alfred had told Grant that the Illuminati was going to hold a convention in Portland, when they would issue orders to their followers as to who would be killed. Alfred said he would be told whether his father was to live or die.

Richard coordinated with the State Hospital and the Oregon court system. It was decided that Alfred would not be committed involuntarily if he served as a test subject in a CAFÉ clinical drug trial. Alfred still believed he might be ordered to kill his father. Grant argued that Alfred stood a better chance of recovering if he received standard care than if he participated in the trial. But the judge overruled him.

***

Alfred had smuggled a box cutter in with his belongings. He decided to kill himself in the shower of his locked unit. He wrote a suicide note and left it on the counter beside the sink.

"Hurting. Good to die."

Alfred turned the shower on full and cut one of his carotid arteries. He dropped to his knees and then laid down. The water washed over him.

***

The Salem police, the Medical Examiner, and the FDA examined the body and the scene of the death. Richard and Edward met with them. The Medical Examiner asked Richard to open the blinding instructions and reveal which medication or placebo Alfred had taken.

Richard said, “The blinding instructions say that Alfred was taking Gabicon.”

***

The next day, Richard and Edward met in Edward’s office.

“As superintendent, I’m supposed to be tough, but I’m sensitive enough to feel that you and I might have helped cause Alfred to kill himself. Do you think we bear some responsibility?”

“Not that I can see. The FDA recently approved Gabicon as an antipsychotic, which means that Alfred could have been prescribed the same drug in standard treatment. There was nothing particularly dangerous about him being in a clinical drug trial.”

Edward said, “Alfred probably felt desperate because he was in a no-win situation; he had to either participate in a drug test or be locked up in solitary confinement. How would you feel if given that choice, especially if you’re sick and feel bad?”

“The mentally ill often have bad choices.”

“But we’re health care professionals. We shouldn’t put a sick person in such a situation.”

Chapter 34: Professor Henry Painter

> Fool me once, shame on you.
> Fool me twice, shame on me.
>
> --Anthony Weldon
> Seventeenth century English politician
> *The Court and Character of King James,* 1617

Ira and Sandra were having coffee in the VA hospital cafeteria.

Ira said, "Dr. Martinson is discovering a lot about mental illnesses."

"He can make a mundane experiment seem like the drug trial of the century."

"Is that a compliment or a criticism?"

"I have mixed feelings about Dr. Martinson. He's accomplishing a lot, but he embellishes his scientific papers."

"What makes you say that?"

"There are rumors."

"Be candid. I won't tell anyone what you say."

"There's a story that he wrote a dishonest paper in 1984."

"What did he write?"

"I was told that he falsely promoted brand-name Thorazine over generic chlorpromazine."

"That doesn't make sense. They're two names for the same drug. They have the same chemical formula."

"You're right. And yet, a nurse told me that many psychiatrists fell for his claim. She said he wrote an unethical paper. I'd like for you to help me determine if he's guilty of malfeasance. That might be impossible for you to do since you admire him."

"I thought something was bothering you."

"I need a fellow whistleblower. I think you have the courage to help me stay with the task if the going gets rough."

"I don't see how you can suspect Dr. Martinson of malfeasance."

"This concerns his interactions with drug companies. Can you analyze his conduct?"

"There's a lot I don't know about Dr. Martinson. "

"Are you willing to help me?"

"If I don't help you, I won't sabotage your efforts."

"Please don't take a position until I've laid out the whole case. And keep this secret."

"You have my word."

"The nurse said Martinson did his 1984 illicit drug analysis in this hospital. He reported that Thorazine was safer and more effective than generic chlorpromazine. In 1985, he gave a speech at the annual meeting of the American Psychiatric Association. He said something had gone wrong in manufacturing chlorpromazine.

"He compared Thorazine to chlorpromazine in an event that supposedly had occurred a year earlier. He said seventeen psychotic patients became agitated when they were switched from Thorazine to generic chlorpromazine. According to Martinson, all the patients in the hospital ward became irritable, and seven became physically combative."

Sandra referred to a note card as she provided more details.

"Martinson said he took the patients off the generic drug. Eight settled down after two days. Six got better after four days. Two recovered to their former state after five days. It took eight days for the last one to recover."

"How did you learn this? You've mentioned a nurse a few times."

"Loretta Getz told me. She helped Martinson care for some of the patients he referred to in his speech. I say 'some of the patients' because only a percentage were real. Loretta said he created the rest of them. She was at the convention and heard the speech. The audience seemed to accept everything he said."

"Why would Dr. Martinson risk getting into trouble by lying to so many therapists?"

"I think the maker of Thorazine, bribed him."

Ira said, "Maybe there's some mistake."

"I want you to examine that possibility."

"Is it okay with you if I find evidence in Dr. Martinson's favor?"

"Yes. If I'm wrong, I don't want to hurt his career."

"What if I refuse to help you?"

"I might look for a different whistleblower partner, or I might go it alone. "

"You feel strongly about this."

"Martinson might still be engaged in wrongdoing. I've heard troubling things about how he's running Pharm Tech."

"I'll look into whether Dr. Martinson lied, but I need you to get documentation for me. I need a copy of his speech and his patients' charts."

"I have those papers. Loretta gave them to me."

Ira said, "If I confirm malfeasance, you and I should report it in a research paper."

"I'd like to write a paper with you."

***

Sandra brought Ira a copy of Richard's speech, as well as hospital records. Ira analyzed attendance records and timelines of the claims that Richard had made.

Ira reported to Sandra what he had found. Again, they were in the hospital's cafeteria.

"The hospital didn't stock Thorazine in 1984. It used generic chlorpromazine exclusively for months. Seven of the seventeen patients who supposedly took Thorazine were not in the hospital at the time. As far as I can determine, they did not exist. Martinson created them and gave them fictitious reactions to the generic drug. So, he didn't just twist the truth. He invented patients and symptoms."

"Anything more?" Sandra said.

"I still can't bring myself to expose Dr. Martinson."

"What would force you to act?"

"I'm afraid that we would not be believed, like Casandra in the Greek myth. She could see the future, but no one would believe her."

"That's a myth," Sandra said. "Give me a real-life example."

"In 1947, Churchill told Prince Bismarck that Hitler would start a war. The German government did nothing to stop that from happening."

"Bismarck shouldn't have been so stupid, but your example doesn't persuade me to stay silent. If we collect evidence, we'll stand a chance of convincing the authorities. If we don't act, we'll be defeated for sure."

"Is there a way we can increase our chance of success?" Ira said.

"Maybe we should get some allies before we go to the FDA."

"I don't want to recruit allies until I'm sure that we're right. We need advice."

"Is there anyone whose word you would believe?"

"I could consult with Dr. Henry Painter at OSU. He's a professor emeritus with credentials in psychiatry, pharmacology, and medical ethics. He taught me and probably Dr. Martinson as well."

Sandra said, "Be careful. He might side with Martinson."

***

Ira phoned Dr. Painter's office, where the professor worked part-time on projects of his choosing. Ira signed up for a meeting.

The meeting day arrived.

"Ira, what can I do for you?"

"I'm considering blowing the whistle on an OSU psychiatry graduate."

"For something done at OSU?"

"No."

"Do I know the person?"

"I believe he was one of your students."

"Don't tell me his name. Let's try to discuss this in an abstract way."

"When I describe his actions, you'll recognize who he is."

"If you identify him, you and I will become liable. You'll be informing on him, and I'll have information that could put me in harm's way."

"I don't have to involve you."

"Over the years, I've told students that they can confide in me. Tell me his name. I'll give you the best advice I can."

"I'll tell you his name before I describe what he may have done. You can tell me to stop talking before I give you information that you might not want to hear."

"Okay."

"Richard Martinson is the person whose honesty I'm questioning. He's chief of psychiatry at Green State Medical College."

Dr. Painter stood up. He looked out the window.

"I bear some responsibility."

"I don't understand," Ira said.

"Do you suspect Richard of doing something seriously wrong? I won't tell you of my experience with him if your concern is over a trivial matter."

"I see it as serious, but I want your opinion in case I'm making too much of it."

"If you're worried about what may be a serious wrong, I have information you need."

Ira described Richard's claim against generic chlorpromazine. He described Richard's speech and technical paper, as well as Sandra's concern that Richard might be falsifying drug tests.

Dr. Painter said, "Richard wrote his thesis on schizophrenia. He gave copies to me and the other two professors on his academic panel. It was a good paper, too good. He had plagiarized much of it. This upset me, but it upset the other professors more. They prepared to expel him.

"I defended him. He apologized to me and the other two professors. I kept him as my student, and he graduated. After hearing of your concerns, I realize that I may have released a con man on society."

"What should I do?" Ira said.

"First, make sure that Richard's speech has not been retracted. If it's been retracted, it will be less harmful, and it may show that Richard has remorse. Tell me who may have been harmed."

"The manufacturer of generic chlorpromazine must have lost sales. Psychiatrists were tricked into thinking generic chlorpromazine is unsafe and ineffective. Some may have quit prescribing it. Patients and insurance carriers paid more for treatment if they quit buying cheap generic drugs. Medicare was harmed if they reimbursed patients for the more expensive Thorazine. If Richard's article was not retracted— as I believe—it continues to mislead researchers and prescribers."

"As a psychiatrist in training, you have an obligation to make sure the article is retracted."

"Just now I've decided, based on what you said that I'll blow the whistle. It seems the only way to limit the harm that Richard may be causing. What should I watch for?"

"Make sure your analysis is correct. If it's wrong, Martinson will crush you professionally. Write the best report you can because it'll be scrutinized. Get a lawyer and go to the FDA."

"Anything else?"

"Watch how Richard's running his clinic. Get back to me if you see anything wrong. Is he publishing only positive results of the drug trials and hiding the negative results?"

Chapter 35: Goofy Theories

> All of this missing data is not simply an abstract academic matter: in the real world of medicine, published evidence is used to make treatment decisions.
>
> --Ben Goldacre
> *Bad Science: How Drug Companies Mislead Doctors and Harm Patients*, 2012

Ira returned to OSU and met with Dr. Painter. The day was sunny. They went for a walk.

Dr. Painter said, "Let's pick up where we left off. Toward the end of our meeting, I wondered how to present topics to you."

"You were concerned about presenting topics to me? With your credentials?"

"Don't sell yourself short. You dropped out of college to earn money to pay tuition. You're not giving up your medical career. You're improving your chances of succeeding.

"You're getting practical experience that most psychiatrists never get. They've never experienced the humility of emptying bedpans and cleaning up puke. And your whistleblowing already is providing you with a grounding that the classroom can't deliver.

"You might end up assessing actions that Martinson may be taking that you're not aware of now. You might learn of problems that you should report to the authorities.

"Many things can go wrong when a researcher cheats in a clinical trial. When a drug trial suggests that the experimental drug is not as effective as a placebo, the drug company might decide to hide the fact. The disappearance of data is a huge problem.

"Richard has published reports that flatter the drug companies. He hasn't published reports of negative data, at least none that I could find. Doctors who read Richard's results are forced to decide which treatment is best without seeing the negative results. This brings up the classic case of an experimental drug called CDX21. Eight subjects suffered terrible reactions. They were put on ventilators because they could barely breathe. Their heads swelled so much they were called elephant men. They lost fingers and toes.

"Richard didn't work with CDX21. So, at first, it seems there wasn't anything he could've done to prevent the harm the drug caused. But when we learn more, it appears that he might have helped psychiatrists avoid the disaster that occurred during the drug trial.

"I learned this possibility from a former student who worked with Richard at the time. Richard had observed severely negative results while studying a similar drug. But he kept the results secret.

"Richard's friend tried to persuade him to publish the negative results, but Richard said a critical paper would damage his relationship with the drug company. If the negative results had been published, they might have saved people from the CDX21 ordeal.

"I suspect Richard still is withholding information. In published data, Fremontria, for example, appears safe and effective. Researchers, however, say that it's barely more effective than a placebo. With its harmful side effects, it does more harm than good. Doctors prescribe drugs based on the balance of all the evidence. Patients are harmed because candid data are left unpublished.

"In medicine, evidence is not just academic. Doctors make life-or-death decisions based on evidence. Context is vital. If I withhold half of the data from you, I probably will convince you of something false. If I flip a coin one hundred times, but tell you of only the results when it lands heads up, I'm misleading you. Researchers are allowed by the FDA to do as many trials as they want and choose which ones to publish.

"Every time we avoid publishing a piece of research, we expose people to avoidable suffering. Sharing results is essential. A systematic review brings together all the evidence on a question. It doesn't cherry-pick which bits of evidence to investigate. Does this drug really work the best of all those possible, or have we been deprived of data? Is there evidence that the drug is dangerous?"

Ira said, "Refusal to publish negative findings is a bigger problem than I thought."

"And we don't know how big. We don't know how much harm it causes because we don't know what the missing data would have shown."

Ira said, "But Richard is just going along with the crowd. Most researchers withhold information. In legal cases, the standard of care is whether the psychiatrist on trial did what other medical professionals would do under similar circumstances. How can I claim that Richard is at fault when he's doing what most others in his field do?"

"That's a big problem that whistleblowers have. Psychiatry is full of goofy theories, like chemical imbalances being a major cause of mental illnesses. The professional guild, the American Psychiatric Association, perpetuates erroneous ideas to sell drugs. When you try to isolate the behavior of one psychiatrist, the fact that many other psychiatrists do the same thing works against you. Richard can argue that he's merely running with the pack. "

Ira said, "So the standard of care is a big part of the problem. The standard is flawed because of the misconceptions about the causes and treatments of mental disorders."

"If you blow the whistle, there'll be times when you have to decide whether psychiatry is stupid or evil or both. Is the APA wrong on the science, or is it lying to help the pharmaceutical industry sell more drugs?"

Ira said, "How does psychiatry know if a treatment works? Maybe the patient would have gotten well on his own. How does anyone really know?"

"Psychiatry gets its answer to that question from clinical trials such as those you're helping Richard do. The idea of using clinical trials for testing hypotheses about mental illness is flawed. Clinical trials are good for evaluating drug treatments for infectious diseases, and it's true that untreated infections, especially those of your eyes and ears, can affect your brain and cause psychosis. But far more mental disorders are caused by sociopsychological factors such as poverty, unemployment, divorce, bullying, and homelessness.

"A randomized clinical trial that requires a main endpoint is a bad way to evaluate psychiatric treatments. We can use some of the same methods to study infectious diseases and mental disorders, but to pretend the two categories will be defeated by the same investigational methods is to delude ourselves."

Ira said, "So we shouldn't use clinical trials to study mental disorders?"

"Only rarely should we use randomized clinical trials to evaluate mental disorders. There are better ways.

"One of the most scientifically unsound things that psychiatrists do is to hold an eight-week trial and pretend that they learned the drug will be a good treatment for years. That's impossible to know from a short study. The most they could surmise is that the drug might be a good treatment for a few weeks. That's the only honest conclusion they can draw. The ability to predict long-term outcomes from short trials is one of the most dishonest claims in mental health.

"The FDA says you have to get your drug approved for marketing. This requirement causes all researchers to do the same thing. The FDA says to test your drug against a placebo, so there's more pack behavior. Use the statistical difference to see if one drug is better than another, and there is more orthodoxy. It's hard for a researcher to have an original thought.

"Researchers do trials until two of the tests make the experimental drug look better than a placebo. They don't have to show a clinical advantage, just a statistical one. When they find a drug's safety and effectiveness are statistically significant when compared to a placebo, it merely means the advantage is probably not due to chance. But researchers claim that the drug works for everyone. And the drug company's marketing department promotes the drug even more vigorously. "

"Can we change this flawed system?" Ira said.

"Scientists must persuade psychiatry to stop worshipping the randomized controlled study as the only way to assess psychotropic drugs."

"I don't think that's possible," Ira said.

"But maybe scientists can get the government to stop using statistical significance as the determinant of symptom reduction. If the symptoms of a mental disorder are reduced statistically, the drug companies claim the drug works, implying that it is good for everyone. A better way is to use the number needed to treat."

"Number needed to treat? Never heard of it."

"The number needed to treat provides evidence for using drugs in selective ways. It indicates that psychotropic drugs should not be given immediately to first-episode patients.

"The number needed to treat causes us to think of the different ways that drugs can be used for different classes of patients. Individuals in drug trials respond differently to the test drug. Some may stay about the same, some may improve a little, some may improve a lot, some may worsen a little, and some may worsen a lot.

"Randomized controlled trials do not provide evidence for using standard doses for everyone. Instead, they provide evidence that prescribers need to develop selective-use protocols. This could identify those who would get well without the drug."

Chapter 36: There's no diagnostic test

> People do not take actions based on logic. We make choices based on emotion. Then we call on logic to justify our choices.
>
> --Kameron Hurley
> *The Mirrow Empire*
> *Worldbreaker Saga,* 2014

Intake specialist Loretta Getz was at Suite Seven, interviewing Melissa McConnell, a 40-year-old prospective test subject, for an upcoming trial of an antidepressant. The trial needed 100 test subjects, 50 control subjects, and 10 healthy volunteers. So far, Pharm Tech had signed up only 30 people.

Melissa said, "I don't know if I'm depressed. I feel blue from time to time."

"I'll take you through the intake form. By the end of the process, we'll know whether you're depressed."

"I can't figure out if I have a chemical imbalance. My understanding is that if I'm depressed, it's due to a chemical imbalance in my brain. Tell me what this condition involves."

"A chemical imbalance means a person's brain has either too little or too much of a chemical called a neurotransmitter. Neurotransmitters are chemical messengers that facilitate communication between neurons in the brain and the nervous system. They influence mental functions, including mood, emotions, stress responses, and thinking. Psychiatrists believe that too much dopamine causes psychosis and that too little serotonin causes depression."

Melissa said, “It’s good to know that psychiatrists are asking the right questions.”

“What I just told you is a fact. It may be hard for a layperson to tell if she has an imbalance. The imbalance makes it hard for you to think clearly. That’s why you need a psychiatrist like we have here at Pharm Tech.

“Here are symptoms of a chemical imbalance. You might be up at night and sleep during the day. You might sleep too much or too little. You’ll feel sad and hopeless. You may feel you’re in danger all the time. You may abuse alcohol or illicit drugs. You’ll have difficulty concentrating. You might avoid other people.”

“Wait a minute,” Melissa said. “These are feelings that anyone might have for a variety of reasons. You just made normal human reactions into mental disorders. The person might have the flu or pneumonia, or maybe a divorce or an eviction. The symptoms you named don’t prove the problem is a chemical imbalance in the brain.”

“You need to have faith in the psychiatrist’s ability to diagnose your problem.”

“I’m not questioning the doctor’s expertise. I’m looking for confirmation from a test. Is there a blood test or a urine test?”

“We don’t have such a test yet, but psychiatry is learning all the time. We’re drawing near a definitive diagnostic test. We may have one within a year. In the meantime, logic helps us make a good diagnosis.”

“Tell me how you use logic.”

"When a psychiatrist lowers the dopamine of a patient, and the psychosis subsides, the psychiatrist figures that lowering the dopamine is worth a try with the next psychotic patient who comes to him. When that second patient also improves, he figures there is even more evidence that lowering dopamine is worth doing.

"Now, the problem with depression seems to be that there is too little serotonin. So, the psychiatrist increases the neurotransmitter."

Melissa said, "Sounds logical to me. Show me where to sign. "

Chapter 37: Prescription Drug User Fee Act

> The Prescription Drug User Fee Act was created by Congress in 1992. It authorizes the FDA to collect fees from companies that produce human drug and biological products. User fees play an important role in expediting the drug approval process.
>
> --*PDUFA Legislation and Background,* U.S. Food and Drug Administration, December 2005

Professor Emeritus Henry Painter and his former student, Ira Cohen, sat on a park bench, immersed in conversation. They were on a part of the medical college campus that was within the Willamette River Greenway. The main stem of the Willamette flowed past them on its northward journey of 175 miles to join the Columbia River. Beginning at Cottage Grove, the Greenway was a 184-mile corridor of water trails, boat landings, footbridges, and parks.

The setting was perfect for the thinking that Painter and Cohen were doing. They were concerned about the health of people diagnosed as mentally ill. They also dwelled on the future of psychiatry. For years, the American Psychiatric Association had said that hypothetical mental disorders should be considered as actually existing in nature.

Ira said, "You've taught me that the drug industry and psychiatry play tricks to make favored drugs look safer and more effective than they are. They design and manipulate drug trials to yield results that indicate the favored drugs are nearly perfect. Which trick is the most effective of the ones that the pro-drug industries use?"

"The dirtiest trick is effective because it's legal, and it makes the FDA complicit. Congress came up with the scheme. Of all of the ways to regulate industries, this is the most corrupt."

"Does this system have a name?"

"The federal law that established the system is called the Prescription Drug User Fee Act."

"If it's so bad, why haven't I heard of it?"

"Because it's legal. And the scheme does result in faster approval of drugs for the marketplace. But there are other ways to get drugs approved without giving the drug companies so much power. The main way would be to give the FDA a bigger budget and let the agency hire more staff. Most new drugs that are approved today are "me too drugs" that cause more harm than good. Fast approval is not always good.

"The PDUFA changes the relationship between the FDA and the drug companies so that the FDA is no longer a true regulatory agency."

Ira said, "Do you have experience with the Prescription Drug User Fee Act?"

"I wrote a book chapter on the topic. The FDA let me participate in meetings on what was then a proposed law."

"Did you get the information you needed?"

"It was 1991 when I started writing the chapter. The Fee Act would be enacted a year later. Consumers, drug companies, and the FDA agreed that it took too long to approve drugs for treating HIV and AIDS.

"The Iverson Company had complained to the FDA about drug approval being too slow. The company had an experimental treatment for AIDS called Chieftan. The CEO, Ron Iverson, accused the FDA of hating homosexuals.

"I remember Iverson saying, 'It's appalling that the FDA delays drug approval because you hate gay men. Dissenters are right to disrupt medical displays at conventions.'

"In 1992, the PDUFA became law," Professor Painter said. "The Iverson Company had filed a formal complaint with the FDA against Dr. David Green, Assistant Chief of the Drug Safety Division. Ron Iverson said he liked the new PDUFA but that the FDA would have to enforce it rigorously if it were to be effective.

"Iverson said, 'The FDA has to be tougher. You've got to come down on employees like David Green, an agency manager. Talk about an insider working against an agency's programs. He's been at cross purposes with the FDA for the 16 years that he's worked for you.'

"Mark Levy, the Chief of Regulatory Affairs, said, 'The Iverson Company filed a formal complaint against Dr. Green. You're now filling in the gaps in your filing. This is your chance to finish making your case.'

"CEO Iverson said, 'The PDUFA will give drug companies the power of the purse. We'll use that power. If we don't like something we're told to do, we won't help pay for it. The PDUFA says you have to negotiate with us. We will no longer be just the regulated party.'

"Dr. Levy said, 'So what did Dr. Green do wrong?'"

"Iverson said, 'He scared people by claiming that Chieftan could make them die of liver cancer. He kept this safe drug from being approved for the market. He got other good drugs pulled off the market after they were approved. He did this through the bad publicity he peddled. Dr. Green said that 62,000 deaths each year are caused by off-label use of atypical psychotropic drugs.'

"Dr. Green said, 'I took my evidence before the FDA. I won most of my cases by doing what epidemiologists do— examining mass markets. Would you have the FDA operate without epidemiologists?'

'I'd have the FDA be more critical of your work and require you to give more credit to the drugs when they reduce symptoms and save lives.'

"Dr. Levy and his panel took a break to consider the testimony they had heard. After two hours, they resumed the hearing.

"Levy said, 'The panel is sympathetic to what Mr. Iverson said. But Dr. Green has civil service protection. Since we have the new PDUFA, we're going to let the new system work for a time to see if it brings harmony to the FDA.'

Dr. Green said, "Congress has made the FDA powerless. Henceforth, the agency will have to play up to companies we used to regulate. Dangerous drugs will be approved even more readily than now."

Chapter 38: Pseudoscience

> If your physician tells *you* that she endorses evidence-based medicine, and that the drug will work for *you* because a RCT has shown that it works, it is time to find a physician who knows that you and the *average* are not the same.
>
> --Angus Deaton and Nancy Cartwright
> *Understanding and Misunderstanding Randomized Controlled Trials,* National Library of Science, 2018

On the medical college campus, a vendor was selling fast food on a public square next to the trail that Ira and Henry were traveling.

Henry said, “I’ll buy.”

They ordered hot dogs, popcorn, and coffee. They walked to the next park bench and ate before they talked again.

Ira said, “I’m getting the impression that people know Dr. Martinson has done bad things but won’t complain.”

“People are afraid the bad guys would remain on the job, and they’d have to either work with or for them.”

“Can’t they overcome those fears?”

“Most people aren’t as brave as you.”

“Is psychiatry in danger of losing its status as a profession? All the lies damage its reputation.”

“Many scientists consider psychiatry a pseudoscience.”

Ira said," What does psychiatry need to do to survive as a profession?"

"Stop lying. Stop creating myths like the one about chemical imbalances."

"What do you think about Thomas Szasz's claim that mental illness is a metaphor?"

"For the most part, he's right. In the 1980 edition of the *Diagnostic and Statistical Manual of Mental Disorders*, psychiatry admitted that nearly all the mental disorders in the manual are hypothetical. They aren't real. They are only what psychiatrists thought was possible. Unfortunately, doctors, regulators, and patients take these so-called disorders to be real."

Ira said, "Are you saying that most mental disorders don't exist in nature?"

"That's right. The 1980 diagnostic manual has 283 so-called disorders in it. Does any genuine scientist honestly believe there are nearly 300 kinds of mental disorders?"

"Are there other psychiatric documents that tell lies?"

"Yes. Rating scales are so subjective that they can't possibly be true for anyone other than the person who was interviewed. Rating scales depend on what a person thinks is true, and every person has a different opinion."

"I'm familiar with the Hamilton Scale of Depression."

"That's a notable example. The Brief Psychiatric Rating Scale and the Panic Disorder Severity Scale are two more. The FDA set the rules for rating scales, and individual psychiatrists took it from there. They created a rating scale for whatever disorder happened to be their specialty."

"What rules did the FDA set up?"

"The 1962 amendments to the Food, Drug, and Cosmetic Act set up the methods by which drug companies were to test whether drugs were safe and effective. These rules said that randomized controlled trials are required and that each trial has to have a primary quantified endpoint. That's why rating scales were invented. Each rating scale produces a number. After you go through the steps called for in the scale, each scale spits out a number that contributes to the primary endpoint. That is how the 1962 dictates are satisfied."

"I didn't realize that."

"Most people don't. They heedlessly follow the experimental protocol for the randomized controlled trial."

Ira said, "People might not use rating scales so much if they knew their history."

"The history of rating scales is contained in the history of randomized controlled trials. Thalidomide was a treatment for morning sickness and other illnesses in Europe in the 1960s. It caused thousands of miscarriages and birth defects. Children were born without legs or arms, or with shortened limbs. In one case, a baby had hands growing from his shoulders.

"Senator Estes Kefauver and Representative Oren Harris led the way to amend the Food, Drug, and Cosmetic Act in 1962. The amendment delivered protection but caused problems that weren't recognized until years later. By then, randomized controlled trials were entrenched as the way to assess the safety and effectiveness of drugs proposed for the marketplace."

"You're saying this is how randomized controlled trials came to be."

"Right. Tony Hill and Louis Lasagna were instrumental in getting randomized controlled trials adopted to prevent another disaster like the one thalidomide caused. In 1962, the FDA began to require drug companies to prove their pharmaceuticals were not only safe but were effective, too.

"The randomized controlled study is a way to assess the efficacy of a treatment for a specific disease. To meet the 1962 efficacy standard, psychiatry developed a model for testing drugs. It developed rating scales to provide a numeric measurement of symptoms of a mental illness.

"Psychiatry wants everyone to believe the short-term reduction of symptoms is evidence of a drug's effectiveness, even if the reduction is barely more than that achieved by a placebo. The drug is called efficacious based on the effects it has on a non-existent average person."

Chapter 39: A Special Project

If you have built castles in the air, your work need not be lost.... Now put the foundations under them.

--Henry David Thoreau
Walden, 1854

It was 1995. Business was booming at Pharm Tech. Drug companies competed for the earliest date that their drug trial could begin.

Arnold hid the money he received from the trials. He was a rich man long before he threw in with Richard. His standing in the community did not change because of the money he received from the trials. He set up shell companies and kept the new wealth out of his name. It would be nearly impossible for authorities to figure out where he hid it.

Richard and Bryce stashed $10 million in cash and securities in US banks and in Barclays Bank in London. Bryce bought a gold Baume wristwatch for $12,000, a diamond ring for $7,000, and gold cufflinks for $1,500. He bought a new home and drove a Mercedes. His share of the profits was half that of Richard's, but it was more money than he'd ever dreamed of having.

***

Richard made an appointment with McQuade and Woods, an architectural firm. A staff architect greeted him.

"I want to meet with your CEO. I have a special project."

"I'm the architect assigned the privilege of welcoming new clients. You'll get the best of service."

"Accommodate me or I'll take my project elsewhere."

"I'll tell the front office. I'll ask for the CEO, but someone will see you right away. Would you like coffee while you wait?"

"Don't keep me waiting."

Within five minutes, a tall man in a black suit entered the room.

"Hello. I'm Charles McQuade, founder and CEO."

Richard shook hands with McQuade.

"I'm Richard Martinson, Director of Psychiatry at Green State Medical College."

"Our chief architect, Jack Maynard, will be here in a moment. I apologize for not being immediately available to you."

"Apology accepted."

Jack Maynard knocked, came in, and closed the door. He introduced himself.

Richard said, "I own 300 acres on both sides of the Willamette. I'll build a castle, complete with a moat and drawbridge. I want the castle to be modeled after one that was owned by Richard the First of early England."

Charles said, "We've been in business for 18 years and never met a project we couldn't handle. Will you live in the castle?"

"Along with my queen and servants. I'll have a moat for security. I'll put sharks in the moat."

"You're joking," Jack said.

"I am not. I want facilities that'll enable me to pour boiling water from the castle wall onto invaders."

"I apologize."

Charles said, "Can you pay a $40,000 retainer?"

"If you can persuade me that you have the ability to handle my project. I'll want you to design the castle, solicit construction bids, help me select the contractor, and manage the construction."

Jack said, "We've designed hospitals, space stations, and amusement parks. We can design a castle, a moat, and a drawbridge. How big do you want the castle to be?"

"Plan on 700,000 square feet for the building itself. A wall will surround the castle, and the moat will be outside the wall. I'll have a courtyard, a barn, stables, horses, a garden, a cow, geese, and chickens. Fields and woods will lie outside the moat."

"You mentioned King Richard. Do you have any particular castle that we should use as a model?" Jack asked.

"The Lionheart owned several castles during his reign. His favorite residence was Château Gaillard, which overlooked the Seine River. I'll pretend that the Willamette River is the Seine. Site my castle so it overlooks the Willamette, like Château Gaillard must have overlooked the Seine. I bought land along the Willamette that has high banks so we can imitate the site of Château Gaillard. Today, the Château is nothing but ruins."

"Have you been thinking about this for a while?" Jack said.

"When I was a boy, I learned of Richard the Lionheart. Dad said I'm Richard's direct descendant."

Richard and Charles negotiated a contract. Richard wrote a check.

The next day, Jack began to research what Château Gaillard must have looked like in King Richard's time.

Richard spent time at McQuade and Woods to make sure the construction plan was perfect. The architectural firm requested and received competitive bids from contractors. Richard selected Northwest Valley Construction.

Jack Maynard worked with Northwest to make sure the construction design was followed. A survey crew staked out the areas where different features would be built. Richard often went on-site with the Northwest Construction foreman.

The foreman said, "The castle will have amenities of modern life, but we'll hide or disguise them so they won't distract from its authenticity."

***

Richard and his girlfriend, Ellen Browning, were at Sotheby's auction house in New York City. They walked in and out of warehouses, looking for any merchandise that would make the castle seem more authentic.

Richard bought a bronze door adorned with a lion's head. He bought twelfth-century tournament armor, a stone lion fountain, a crystal chandelier, and a coat of arms. He bought paintings, marble vases, palace rugs, antique furniture, and pendants.

Richard spotted two huge concrete lions. A sign said the bidding would start at $20,000.

Richard said, "These lions will warn people to beware when they approach the castle. Folks will see the lions before they cross the moat."

"Besides the purchase price, you'll have to pay for transporting and storing the lions."

"But this might be my only chance to get statues that will so effectively warn against attacking the Lionheart."

"They're about to be auctioned."

Richard had presented a credit card when he had first qualified as a bidder. Now, he showed a yellow, numbered paddle when he made a bid.

Richard's first bid was for $22,000. His next one was for $26,000. The final bidders were Richard and a bearded man. Finally, the man quit bidding, and Richard bought the lions for $51,000.

A Sotheby's contractor asked Richard for the job of transporting the lions to River Town. They settled on a price.

Chapter 40: The Geese

> The temple of Juno kept geese, which were considered sacred. While the Romans slept, Gallic soldiers climbed to the top of the cliff that protected the seventh hill of Rome. The geese honked and woke up the Roman soldiers who repulsed the Gauls. Thus, geese saved the city.
>
> Paraphrased from Livy, *History*, Book V, Chapters 47-49 Titus Livius, 57 BC-AD 17

Northwest Valley Construction built the castle, the surrounding wall, the barn, and the stables. The company dug the moat outside the wall and lined it with riprap.

Last of all, Northwest installed the drawbridge and the entry lions. Chains rattled when the drawbridge was lowered.

Richard and Ellen moved into the castle. They lived like royalty, wearing robes and crowns. Richard called Ellen "my queen." She called him "Lionheart."

They invited Bryce to see the castle. He stifled laughter when he saw the couple in their royal wardrobes.

Richard hired four servants and provided them with clothes characteristic of early England. Besides cooking, the servants took care of the castle, the premises, the garden, and the farm animals.

During Bryce's visit, Richard released three dozen geese into the courtyard.

"Why so many geese?" Bryce said.

"In Rome in 390 BC, geese were kept in the temple of a fort. According to legend, Juno, a Roman goddess, considered them sacred. One night, Gaelic troops attacked the fort. The geese woke the Romans. It wasn't the human guards or the dogs. Geese sounded the alarm."

"How do you know that?"

"Livy, an ancient Roman historian, wrote an account. I bought Roman Tufted geese, the species that saved the Romans."

"Do you expect to be raided?"

"It's possible."

"Are you taking other security measures?"

"My servants practice with swords, long bows, and crossbows. And we have oil and boiling water that we can pour on invaders."

"What about modern weapons?"

"We may need to defend a castle, not overtake a kingdom."

Chapter 41: The Airbourne Wheel

> Every country needs its whistleblowers. They are crucial to a healthy society.
>
> The employee who, in the public interest, has the independence of judgment and the personal courage to challenge malpractice is a public hero.
>
> --Fuad Alakbarov
> Human rights activist. Born 1988

Ira and Sandra met in the medical college library. Ira told Sandra what he had learned from Professor Painter.

"So, you decided to report Martinson's misconduct?"

"Evidence shows he lied about generic Chlorpromazine. I will report that to the FDA."

"Instead of saying 'I,' try saying 'we.' Include me in our plans."

"I apologize. I'm used to speaking only for myself."

"Well, I'm the one who told you of Martinson's malfeasance. You've got a partner."

Ira said, "We're in this together."

"Tell me more about what you learned from Professor Painter."

"Maybe you and I should resign after we report to the FDA. Whistleblowers almost always suffer retaliation from their employer after they report malfeasance."

Sandra said, "An employer might fire whistleblowers or assign them unpleasant jobs. Punishment also includes low-performance ratings and demotions. But if we worry about retaliation, we won't be effective at blowing the whistle. First, let's go to the FDA."

"I'll make an appointment."

***

Ira and Sandra drove to Salem. They met with the FDA compliance officers Ted Watson and Nelson Hubble. The whistleblowers described how Martinson had discredited generic Chlorpromazine. Ted asked Ira and Sandra to wait in the lobby.

Nelson phoned a SmithKline office in Philadelphia and spoke with Perry Wertheimer, the Vice President of Marketing in the United States.

"Ted and I are meeting with two whistleblowers from a clinic run by Green State Medical College. We sent them to wait in the lobby."

"What do they say?"

Nelson described Ira and Sandra's accusations.

"Keep them waiting. I'll call you back."

They hung up. Twenty minutes later, Perry called back.

Perry said, "I know Martinson. I'll tell him that his employees are blowing the whistle. He'll fire them. That's SmithKline's position.

Tell the whistleblowers that you're not interested in what they've said."

Nelson said, "We have to be careful. Porter and Cohen might go to the news media. Reporters might identify with these young people. We don't want to cause reporters to expand this story."

Ted asked Ira and Sandra to return to the meeting room.

Nelson said, "You say that Dr. Martinson made his speech in 1985?"

"That's right," Ira said.

"That's five years ago. And this is a small issue. Do you have any idea how many more important issues we deal with every day?"

Ira said, "Martinson's article remains in the scientific literature. It influences which drugs are prescribed. The FDA should have told the publisher to retract it."

Ted said, "We're not going to do anything with your informing on Dr. Martinson. He could fire you for meddling in his research."

"I can't believe this," Sandra said. "We've given you copies of the treatment records, Martinson's speech, and the published paper. You must investigate."

"Don't tell us what we must do," Nelson said. "You might have stolen medical records. Drop this, or we'll report you to Dr. Martinson and to the VA Hospital."

Ira said, "We withdraw our complaint."

Ira and Sandra returned to their car.

Sandra said, " I never dreamed this could happen. Those guys are covering for Martinson."

"Let's drop this case. Let's switch to Martinson's current wrongdoing. We can get evidence and go to a US Attorney instead of the FDA. We'll come up with a case that can't be brushed aside."

"Will the FDA tell Martinson that we reported his wrongdoing?" Sandra said.

"The FDA wants to look impartial. Ted and Nelson will urge SmithKline to report us to Martinson. It'll be the drug maker who tells on us."

***

At SmithKline headquarters, Perry Wertheimer told his supervisor that he would notify Richard Martinson of what Ira Cohen and Sandra Porter claimed. He would make the phone call the next morning.

Traffic was heavy when Perry climbed into his company's limousine. He relaxed with a whisky sour.

Mitchell, the chauffeur, said, "How was your day, Sir?"

"The FDA said whistleblowers lied about a research organization that does clinical drug trials for us. I'll get the squealers fired. I'll make the call first thing in the morning."

"Do you have to do a lot of that kind of thing?"

"Yeah. I do."

A truck ahead lost a wheel. It flew through the air, bounced, and smashed into the front of the limousine.

The motor car flipped and skidded on its top.

Mitchell suffered internal injuries and broken bones. Perry was killed.

Chapter 42: The US Attorney

> The most misleading assumptions are the ones you don't know you're making.
>
> — Douglas Adams
> English author, 1952-2001

Sandra and Ira were in a pizza parlor just off campus. They had eaten and were finishing their beer.

Sandra said, "How do you feel since we failed at blowing the whistle on Martinson?"

"If SmithKline told Pharm Tech of the attempt we made, the word didn't get to Bryce. If he were told, he would have fired us."

"I can't imagine why SmithKline wouldn't have told on us."

"Those FDA stooges put me in a fighting mood," Ira said. "I'm determined to reduce the harm that Pharm Tech causes."

"How much risk ought to be allowed in a drug trial?" Sandra said. "The risks seem too great for the benefits that most test subjects receive. I think most subjects assume the risk is worth it because the test is done to acquire scientific knowledge. I wonder what they would do if they realized the real purpose is no more than putting a drug in a better marketing position.

"How much risk is justified in a study whose aim is to produce commercially attractive messages? Regulators should be concerned that the Nuremberg Code said the purpose of the study has to be of great importance to justify experimentation on humans."

Ira said, “It’s time you and I reported research fraud at Pharm Tech. The clinic tested ineligible subjects .”

“I’ve written a statement about Richard and Bryce bribing me to not report Sam Winegart’s suicide attempt. I admitted that I accepted their offer after refusing it twice. I also documented that Bryce destroyed Sabrena Southwick‘s medical records and hid the fact that she had liver cancer.”

“If we want to qualify for qui tam funds, we need to report the fraud to a federal agency,” Ira said. “Since we can’t trust the FDA, our only choice is to go to the U.S. Attorney.”

***

Three days later, Sandra and Ira met with U.S. Attorney Carlos Ramirez in his office in Salem. Ira handed him a file on Martinson and Jewel. Ramirez read it.

After ten minutes, Ramirez spoke.

“You’re reporting fraudulent drug testing but not theft.”

“Theft?” Ira said.

“You’re concerned about the welfare of test subjects, but you didn’t mention theft.”

“What do you see that we don’t?” Sandra said.

“Why would Martinson and Jewel run so many drug trials?”

Sandra said, “To find treatments for mental disorders, to publish scientific articles, and to get good performance ratings.”

"Do they get anything more?"

"They collect a fee for each person they test," Ira said. "But that goes to the medical college to fund Pharm Tech. It doesn't go to Martinson and Jewel."

"Who owns Pharm Tech? "

"The medical college," Sandra said.

"Have you seen the deed?"

Sandra said, "The college pays our wages."

"Maybe so, but does the college own Pharm Tech? Could Martinson and Jewel be the owners?"

"What makes you ask that?" Sandra said.

"How much money does a drug company pay Pharm Tech for each subject tested?"

Sandra said, "Bryce said that Pharm Tech gets $2,000 for each schizophrenia patient that it tests."

"The figure is more like $20,000. That much money can cause contractors to do illegal things."

"Newspaper ads that Pharm Tech runs bear the college's logo," Sandra said. "The company pays our wages, and we use college stationery and business cards."

"None of that proves who owns Pharm Tech. Ownership would explain why Martinson and Jewel do so many trials."

"What do we do now?" Sandra said.

"I'll get a court order that'll authorize me to examine any Pharm Tech bank account. I'll get answers. If you tell anyone what I've said, I'll prevent you from getting qui tam funds. Pledge to me that you'll not breathe a word of this until after the case has ended."

Ira said, "We won't say a word."

Sandra said, "I agree."

Ira said, "How could Martinson and Jewel own Pharm Tech without the college knowing?"

"The college probably owns a corporation that handles a few drug trials, but more trials are done by Pharm Tech, which I believe belongs to Martinson and Jewel. They might have put the deed into the name of a corporation they own."

"How could they do that?" Ira said.

"A college big shot probably is protecting them."

"So, these dishonest drug deals may include larceny," Sandra said.

"Grand larceny," Carlos said.

***

Carlos used the file that the whistleblowers had given him to get a court order that authorized him to examine Pharm Tech's bank accounts. The main account showed big deposits and withdrawals.

Carlos put off reporting the apparent fraud to regulators because he had more investigation to do.

Chapter 43: The City of Roses

No great mind ever existed without some touch of madness.

--Aristotle
Ancient Greek philosopher

It was January 1995. Richard and Bryce had built their business by rapidly recruiting test subjects and by delivering favorable results for drug companies. Now they needed to recruit test subjects even faster to keep up with the increased demand for their trials.

The partners had eaten dinner at Spring Valley Restaurant in River Town. They sipped wine and talked about their business.

Richard said, "River Town can't meet our need for test subjects. We need to recruit in an additional city."

"The City of Roses has a huge homeless population."

"That's where we need to go."

Bryce said, "Would we test our recruits in Portland?"

"We'd bring them to River Town, where we have facilities, equipment, and staff."

"But we'd need some sort of facility in Portland. We'd need to do a preliminary screening there."

"I agree. We'd need to screen the prospects enough to judge whether they were likely to qualify as test subjects. We'll need to do a preliminary, cursory screening. Otherwise, we'll waste money bringing unqualified subjects here."

"How will we staff the operation?" Bryce said.

"We'll use a contract research organization. I'd like you to set up the operation."

"Will you supervise our trials here while I'm in Portland?"

"Of course."

***

Martinson hired Quantum Medical Researchers to manage the new office. Quantum assigned the task to John Spores, a staff psychiatrist.

Bryce spent three days a week at the new facility. He forged Richard's name on papers saying that Pharm Tech had given each prospective test subject a physical exam.

John Spores knocked at Bryce's office door. Bryce invited him in.

"Are you a physician? People call you Dr. Jewel."

"I'm a PhD pharmacologist. I took psychiatric courses at OSU. And I've learned a lot more on the job."

"I object to your signing Dr. Martinson's name on medical papers. It's a forgery to sign someone else's name, no matter what the agreement is between you and him."

"You don't need to be concerned. You're not forging anything. You won't be accused of doing anything wrong."

“That might depend on how serious a legal charge might be. If the charge is a felony and the offense is ongoing, a judge might decide that I’m complicit.”

“Don’t jump to conclusions. I sign Richard’s name to save us a lot of trouble. It makes our business more efficient. If a matter is complex medically, Richard makes the decision.”

“Well, I don’t like it, and the FDA might not either.”

John walked out of the office. Bryce phoned the Quantum headquarters in Seattle. He spoke to John’s boss, Irma Casselbaum.

“I want you to replace John Spores.”

“He’s one of my best employees.”

“He’s too concerned with Pharm Tech internal matters. He shouldn’t care about how my boss and I interact.”

“Unless it affects how you recruit test subjects.”

“I’m not going to argue with you.”

“Tell me what he did wrong. I don’t want his replacement to make the same mistake.”

“Sometimes I sign my partner’s name. That’s none of Quantum’s business.”

“I’ll tell John to pack up.”

“Who’ll replace him?”

"Erick Taggart. He's perfect for the job."

That same day, Irma called Erick into her office.

"I want you to take charge of a recruitment office in Portland. Are you up to it?"

"Recruitment is what I do."

"The manager you're replacing objected to our client pretending he's a physician. We don't have to make the people we work for follow the letter of the law. So be flexible."

"Flexible is my middle name."

***

Erick Taggart didn't object to Bryce's business practices. Bryce came to trust him.

Bryce needed someone to confide in. He had no wife. He could not confide in Richard, for he was the main reason for Bryce's concern. He couldn't share secrets with his staff because they might report him for any improper activity.

Bryce worked to make friends with Erick. They were having lunch at the Spaghetti Factory.

"You're more flexible than your predecessor."

"It's none of my business how you and your partner share work. My job is to help you recruit people for your clinical trials."

"I want to explain what happened between John Spores and me. I wanted Quantum to help me. I didn't want an overseer. I can understand if you don't trust me after I got John replaced, but I wrote a letter recommending him as a manager. I have to sign for Richard because he's gone so much of the time. He's Pharm Tech's only psychiatrist. We have three psychologists, but by law, they can't prescribe medication."

"I understand," Erick said.

The conversation drifted onto neutral subjects. Bryce told a joke, then Erick told one.

"I need advice," Bryce said.

"I'm not Ann Landers."

"I sense that you blame me for having Irma replace John. Richard is rarely around, especially now that he has a castle where he writes research papers."

"He bought a castle?"

"He had it built."

"Why would he do that?"

"He admires ancient kings."

"I realize I'm being drawn in, but go ahead and tell me about your strange partner."

"I've known Richard since we were farm kids. In our games, he was Richard the Lionheart. His dad said he named Richard after

King Richard the First. I think Ethan made that up to make Richard feel important.

"We were playing a game in which Richard was Lionheart and I was Robin Hood. He fell on a stairway and injured his head. Now, sometimes he seems to think he really is the ancient king. I can't predict when the delusion will show up. We've gone years without the king appearing. But then there will be episodes in which he reigns supreme."

"Sounds serious."

"It must have cost millions to have that castle built. Richard treats his girlfriend as his queen. He stocked the castle with staples like potatoes and squash. He has a garden, cattle, geese, and chickens. And he's training archers to defend the place. A wall and a moat surrounds the place. I believe he'll leave the college and hide out in the castle."

"Could Pharm Tech operate if Richard bails out?"

"No. He's Pharm Tech's mastermind."

"When did you realize that his weird belief might bring Pharm Tech to an end?"

"When John Spores showed me that my forgeries are obvious. He made me realize that I can't keep filling in for Richard, especially if he gets more delusional. I've decided to leave Pharm Tech, and I need help in making the break."

"You've got drug companies desperate for you to do clinical trials. Until I worked here for several days, I had no idea how big your clientele is. You'd be walking away from a fortune."

"But I believe authorities will be tipped off. Or Richard might go mad, and we could get caught that way. I heard him talking to himself in his office. He was talking about the Third Crusade. Once, he said he is like two people. That's when I decided to get out while I can."

Erick looked dumbfounded. Bryce resumed talking.

"Richard said he was taking cues from the pharmaceutical industry, which often breaks the law and operates in plain sight of the authorities. He doesn't plan to escape. He thinks he can stay in his castle and defend it against law enforcement."

"Amazing," Erick said. "The guy is full of tricks. That's why you two made so much money. Now he plans to keep his fortune by imitating the drug industry. His trickery reminds me of the old shell game."

"Hiding a pea under a walnut shell. You're right."

Erick said, "I can see why this is rough on you, but I'm not in the advice business."

"I want to live in another country. I want help in relocating and keeping authorities from learning where I go. After I leave, Pharm Tech will fail. And Richard probably will try to blame me for running illegal drug trials."

Tell me exactly what you want," Erick said.

"I need to change my identity. I need help in getting passports, airline reservations, hotel room reservations, and a place to live. I could do these things myself if they didn't have to be done in secret. I'll pay for the help. All you have to do is find me someone who does this kind of work."

"You'll have state and federal cops after you."

"I have no idea how to find a fixer."

"What makes you think I do?"

"You're a man of the world. When I told Irma I needed someone to replace John Spores, she sent me you. "

"Okay. I'll find a fixer for you, but it'll cost you, and it'll take a while."

"I don't have time. The authorities could get wise to Pharm Tech at any moment. I need the name of a fixer in two days."

"The pressure on me to hurry will cost you more."

"What's the bottom line?"

"$10,000."

"Okay," Bryce said.

"Half now. Half when I give you the fixer's name. I won't guarantee results, only that he does this kind of work and is willing to help you."

Bryce and Erick went to their office. Bryce paid his confidant $5,000 in one-hundred-dollar bills.

Two days later, Erick came up with a name, Conrad Reynolds, and received Bryce's second $5,000.

***

Bryce and Conrad met. They agreed that Bryce would pay $50,000 and would deliver half immediately. Bryce paid the fixer $25,000 in cash.

Two weeks went by. Bryce finally received the phone call he'd been waiting for. Bryce and Conrad met in a rented room in the Portland Hilton. Conrad handed his confederate a packet of forms.

"Here are six documents. In a week, as James Chase, you'll board a plane to London. The first document in your packet is your airline ticket. The second item is your passport for Chase.

"You'll stay overnight at the Clarion Inn. The third item is a room reservation for Chase. The fourth item is an airline ticket from London to Sydney. The fifth document is a passport that enables Chase to enter Sydney from London. The passport includes a visa allowing Chase to spend more than fourteen days in Sydney. You'll stay two nights at the Wild Swan Inn, so the room reservation number is the sixth item in your packet.

"From there on, you'll be on your own. I didn't find you a residence in Sydney. I don't want to know where you will live. I want to be shed of you as much as you want to be shed of your present situation. Erick didn't tell me why you want to leave River Town, and I don't want to know."

Conrad watched as Bryce counted out $25,000 in one-hundred-dollar bills. He went on his way.

***

Bryce sold his home at a low price. He got rid of his car two days before he left for London. He sold it for half price to the dealer where he had bought it.

Bryce made sure things were operating smoothly at the VA Hospital, Suite Seven, and the Portland recruitment office.

Chapter 44: I drifted off for a second

When the jig is up, there's no need for more dancing.

--Earl Derr Biggers
1884-1933
American novelist. Creator of Charlie Chan.

An assistant of Carlos checked with the banks in River Town, showing them the federal court order. A vice president at the First National Bank immediately told the assistant US Attorney of the Fairway Industries account.

Carlos compared the Fairview Industries account to the medical college's account that was used for drug trials. The Fairview bank statement showed more activity and bigger balances. The college account balance had shrunk as soon the Fairview account opened.

Carlos studied the college's organizational chart. He looked for which administrator appeared most likely to protect Martinson and Jewel as they took money that belonged to the college.

He phoned the campus police. A day later, Carlos met with Captain Tommy Wong in Wong's office.

"This is my first meeting with a U.S. Attorney. How can I help you?"

"First, assure me you'll keep secret everything we discuss."

"I will."

"How long have you been with the campus police?"

"Since 1978."

"So, you know things about the faculty and the staff?"

"Yes."

"It would be a federal crime to tip off anyone about what I'm about to say."

"I won't tell anyone anything."

"If you're friends with anyone whose name comes up, tell me at once so I can decide whether to say more. Are you friendly with anyone working in financial management at the college?"

"I know the managers. We don't socialize."

"Does the name Pharm Tech mean anything to you?"

"That's a drug-testing clinic that the college owns."

"Pharm Tech is why I'm here. It might be involved in medical fraud. Do you know of anything suspicious about the clinic?"

"No."

"If there's fraud, who in the college administration is in the best position to cover up for the Pharm Tech operators?"

"Pharm Tech tests drugs for pharmaceutical companies. Is someone keeping the money for himself?"

"It's worse than that. The college doesn't own Pharm Tech. The college owns a clinic that's barely active. The wrongdoers use the college's clinic just enough to camouflage their illegal acts. College employees created a new clinic that they named Pharm Tech."

"I think you got your facts wrong. I know a coordinator who works there, and she's paid by the college. The college owns Pharm Tech. Everyone knows that."

"Everyone assumes the college owns Pharm Tech. But they don't realize there are two clinics. The original clinic operates out of the VA hospital. It has never had a building of its own. Since Pharm Tech leases a building, called Suite Seven, it's easy to mistake it for being the only clinic. "

"That's hard to believe."

"The best rackets fool everyone. Who in the college's financial structure might protect the faculty members who are running Pharm Tech?"

"The college's chief financial officer is Arnold Picket. He's in the best position to protect a swindler. I'm not accusing him. "

"I'll interview Picket, but I don't want him to see it coming."

"I have a feeling this will involve me."

"I want you to schedule a meeting for yourself. I'll go along and then interview him. He'll be caught off guard."

"That could ruin my relationship with him and maybe other administrators. I might be seen as a traitor to the college."

"You're a sworn law enforcement officer. I'm the U.S. Attorney for this district, and I have evidence that state and federal laws are being violated. Money is being diverted from your employer. Your bosses will be upset if you don't help me."

“To sneak you into an interview isn’t one of my responsibilities.”

“What are you willing to do?”

“I’ll tell Arnold’s secretary that you have concerns. If she asks questions, I’ll say it’s best for me to let you speak for yourself. That might make your job harder, but we’ll be on ethical high ground.”

“Are you willing to make the call now?”

‘Yes.”

Wong called Picket’s office. He said the U.S. Attorney wanted a meeting involving federal law. Picket’s secretary did not ask for details.

The next day, Ramirez and Wong were in Picket’s office.

“I hope the college didn’t violate Title IX again,” Arnold said. “Maybe you should be talking to our athletic department.”

Carlos said, “This involves Pharm Tech, a drug-testing clinic. The way it’s being run may be in violation of federal law.”

“Pharm Tech has always been above board. Why are you talking to me? I have never interacted with the FDA. The main federal agency that I work with is the IRS. We paid our taxes. I can show you the records.”

“Do you know Richard Martinson, the psychiatrist who runs Pharm Tech?”

“Yes.”

"How well?"

"College activities require me to interact with him."

"Have you ever interacted with him alone?"

'No."

"Do you know Bryce Jewel?"

"No."

"Do you know who he is?"

"No."

"What do you know about Pharm Tech?"

"It's a college clinic with an excellent reputation."

"What do you know about the clinic, what it does, and so on?"

"Pharm Tech runs clinical drug trials for pharmaceutical companies. It does the trials at the VA hospital under a cooperative agreement between the college and the VA."

"Where does Pharm Tech get its funding?"

"From companies whose drugs Pharm Tech tests for market approval by the FDA. The VA hospital gets some of the funds because it provides hospital wards and a laboratory."

"Do you know when Martinson began to run the trials?"

"It's been a few years."

"Did drug testing increase during the time Martinson has run Pharm Tech?"

"I imagine there's been an increase. Nationwide interest in psychotropic drugs has grown in recent years."

"Does Pharm Tech use the on-campus institutional review board?"

"I assume so."

"Have you seen the IRB's records concerning Pharm Tech?"

"No."

"Have you seen Pharm Tech's bank account?"

"No."

"Is your organization responsible for monitoring Pharm Tech's bank account?"

"One of my accountants does that."

"Which one?"

"Do you want me to check my records for the name?"

"No. I'll give you a list of questions to answer."

"Why are you asking me these questions?"

"I received a complaint that Pharm Tech is diverting money from the college to private individuals."

"I've never heard anything like that."

"This may have gone on for years."

"I can't help you."

Carlos and Tommy thanked Arnold and left his office."

***

Arnold left work early. He phoned Richard at home.

"I'm at the Yellow Canary. Can you meet me here?"

"I'll meet you at The Mad Greek in an hour."

Arnold arrived first. He sat at a table in the back of the dining room. Richard found him. Arnold described the U.S. Attorney's interrogation.

"I didn't expect to be grilled by a federal cop who apparently has already examined Pharm Tech's bank account. How did he get wise to us? What have you got us into?"

"I made you more money than you get from your job and investments, too."

Arnold said, "And you might have got me a prison sentence too."

"What do we do now? Do you think your answers satisfied Ramirez? Will he back off? Maybe he's too busy with other cases to bother with us."

"He seems determined to get us," Arnold said. "He's got the head campus cop stirred up. We'd better stop doing anything illegal."

"Do we stop using the commercial IRB that Bryce found? Do we decide we've made enough money? Should I tell the drug companies to stop sending deposits to my home?"

"Wait a minute. Are drug companies sending trial fees to your home instead of to the medical college? Did I hear you correctly?"

"Well, yes," Richard said. "You know that."

"No, I don't. That's a stupid thing to do. You never told me that you were going to do anything like that. That's brazen and stupid."

"You approved of me having drug payments sent to my home."

"You never consulted me about turning your home into a deposit drop box. I would never have approved such a move. I'm done with you."

"I have more questions for you."

"To hell with your questions. I didn't think you were doing anything as stupid as taking payments at home. That's practically admitting that you're on the take. I'm pulling out of your operation. I'll insist that we aren't partners. I can ride this out. If you blame me, I'll tell the authorities how you cooked up the whole scheme. My lawyers will take after you."

"At least Jewel is standing by me. He's a faithful subject."

"Subject? He's not a test subject. He's your employee."

"He's loyal to me. I'm his king, and he's my subject."

"What are you talking about? Have you gone nuts? Has the stress got to you?"

"Forget it. I drifted off for a moment."

***

The next morning, Richard went to Suite Seven. None of the staff had heard from Bryce. Richard tried to phone him at his home. A recorded message said Bryce would return the phone call.

Over time, Richard had withdrawn money from the Pharm Tech bank account and put it in secret overseas accounts. Now he was reaping the benefit of that strategy. There was no need to go to the bank to make a withdrawal. He already had most of the money.

Richard went to a supermarket and loaded his car with groceries. He drove to Château Guillard and carried in his purchases.

"Ellen, I'd like you to go to town with me to buy supplies."

"I went to the store yesterday. And you just now carried in some stuff."

"Things have changed. We may be hemmed in for several months."

"What's going on? You're scaring me."

"It has to do with Pharm Tech."

"I can't be cooped up without knowing why."

"I'm in trouble with the FDA."

"I knew you'd get in trouble sooner or later."

"I need you to stay with me. I'm King Richard transplanted into a modern setting. I hide my identity from everyone but you. You're my queen, and I need you to stay with me as I fight a battle as big as the Third Crusade."

"Long ago, I accepted you as Lionheart. I'll stay with you for now."

***

Richard and Ellen made another trip into River Town. They bought food for their livestock and filled their gas tank.

They returned home at 11 p.m. Their headlights shone on the entry lions and the raised drawbridge. Richard pressed a combination of buttons on his handheld garage door opener. The drawbridge came down and settled in place.

Richard drove into the courtyard. He raised the drawbridge and parked in the garage. He and Ellen took the groceries into the castle.

Chapter 45: Bryce and Richard are missing

There is no honor among thieves.

Cicero (108-83 BC) was one of the first to express the idea.

James Chase bought a condo on the east edge of Sydney. From his balcony, he could see the Tasman Sea.

James strolled through Sydney, taking in the sights. He ate in restaurants and fixed only tea or coffee at home. He kept to himself.

***

Some of the test subjects that Jewel and Taggart had sent to the VA hospital did poorly. Forty-two-year-old Wendel Franklin killed himself after taking Fremontria for three days. He cut one of his carotid arteries with a pen knife he had smuggled into the hospital.

Wendel's brother, Sherman Franklin, went to Suite Seven the day he learned of Wendel's suicide.

"I'd like to see Dr. Martinson. I'm Wendel Franklin's only living relative."

"I'm sorry, the doctor's not here," Loretta said.

"How about Dr. Jewel?"

"He's not here either."

"Who's in charge?"

"I am for now."

"Have you tried to contact Martinson?"

"Yes, without success."

"Have you tried to contact Jewel?"

"Yes, also without success. Everyone at Pharm Tech is sorry about what happened to Wendel."

"So, who's the next doctor in line? Who do I talk with?"

"No replacement has been assigned just yet. This all happened so fast."

"What does the college make of Martinson and Jewel being absent?"

"The President's office is scheduled to come over this afternoon. They'll put someone in charge. They're coordinating with the FDA."

"What if neither Martinson nor Jewel shows up?"

"The clinical trials that are underway are our biggest concern. VA doctors are assuming responsibility for the short term."

Sherman said, "It looks like Martinson and Jewel skipped out."

***

Sherman went to the FDA offices in Portland. He demanded that the agency shut down Pharm Tech. He was told an investigation was underway.

The news media reported that Martinson was in his castle and that the police had ordered him to surrender. The media also said it appeared Jewel had fled the country.

Sherman Franklin offered a $100,000 reward for information about Jewel's whereabouts, to be paid as soon as the fugitive was found guilty.

****

Erick Taggart got off a Portland city bus and walked to his apartment. At the door, he punched in a code that let him enter the building. He rode the elevator up to his unit. *The Oregonian* lay near his door.

He read a story about Richard Martinson hiding out in his castle and about Bryce Jewel disappearing. The story also told of Sherman Franklin offering a $100,000 reward for information leading to the sentencing of Jewel.

Erick phoned Conrad Reynolds.

"Have you heard about Wendel Franklin's suicide?"

"I heard."

"What else have you heard?"

"Wendel's brother is offering a $100,000 reward for information that gets Jewel tried and convicted."

"Can you meet me for dinner?"

"The Oyster Bar on Ankeny Street?"

They met at the restaurant. After eating, they discussed the news about the former proprietors of Pharm Tech.

Conrad asked, "Will Quantum lose Pharm Tech as a client? It seems that at least the clinic's drug testing will dry up."

"My boss said authorities are investigating whether Quantum and Pharm Tech took a suicidal man into a clinical drug trial. We did what the FDA requires and what drug companies do. I'm not worried."

"Why did you invite me to dinner? as if I didn't know," Conrad said.

"I'm interested in the $100,000 reward that Wendel's brother is offering."

"That's a lot of money."

"It sure is," Erick said.

"You think I know where he is."

"Do you?"

"I could go to prison if I admitted I helped him flee the country."

"You need a mediator," Erick said.

"I bet you have someone in mind."

"I can see that you get most of the reward without either of us getting in legal trouble. The cops will look like heroes if they arrest Jewel, and they can't get him without our help. He could be anywhere in the world."

"Whether the cops will work with us depends on how much they want him," Conrad said.

"They want him. They want him to testify against Martinson. They want the glory of catching him."

"How much money do you want for negotiating a deal?"

"How much do you need?"

"$80,000."

"The reward is $100,000. If you get $80,000, that leaves me with only $20,000. I won't deal with the cops for only $20,000."

"What do you want?"

"$30,000," Erick said.

"I'll agree to you taking $30,000 if you make a deal that protects me from prosecution. But it's got to be an air-tight deal in writing."

"I won't give the cops your name. If they don't know who you are, they can't arrest you. They'll deal entirely with me. I'll pay you as soon as I get the money."

"How will you make sure the cops don't follow you or monitor your phone calls?"

"Let's arrange where and when you and I will meet, so we don't have to make phone calls that might be intercepted. And I have ways to lose anyone who tries to follow me. I'll make sure the cops don't bug me. I'll search through my clothes and will shower after I meet with the cops. I'll warn them that any attempt to bug me will kill the deal."

"Are you willing to not do the deal if the cops want to arrest me? They might even put you in jail."

"Sure," Erick said, "I believe the cops would do a deal with the devil to catch Bryce. They have a terrible public image right now. Tell me where he is."

"I'll tell you as soon as you make a deal that protects me."

Chapter 46: Let's make a deal

I didn't say it was your fault. I said
I was going to blame it on you.

-- Ashleigh Brilliant
American author and cartoonist
Born December 1933

Erick phoned the River Town Police Department. He told Lieutenant John Hanlon that he could help the police find Bryce Jewel. He asked Hanlon to meet him at The Mad Greek Restaurant.

Erick sat at a table at the back of the dining room. He stood and introduced himself when Hanlon and a man in plain clothes found him.

Hanlon said,"This is Detective Alex Emory. We'll record this conversation."

Hanlon set a cassette recorder on the table and turned it on. Erick set a recorder next to Hanlon's and turned it on.

"You said you know how to find Bryce Jewel. How is it that you have this information?"

"Before I answer that question, I want you to answer one: "Do you want to make a deal? Does my man qualify for the reward offered by Sherman Franklin?"

Emory said, "Your question is moot because we're about to force you to identify your man."

"Unless you work with me, you'll never find Jewel. Acting through me, my man will give me the information that you want. If you arrest me, I'll not tell you who the would-be informant is. Franklin will be upset with you guys for losing the chance to bring Jewel to justice. And you'll look even worse to the public than you do now."

Hanlon and Emory looked at each other.

Hanlon said, "Stand up and place your hands behind your back."

The officers handcuffed Erick and took him to the River Town jail. They locked him in a cell. Two hours later, a jailer got Erick out of his cell and took him to an interrogation room where Hanlon and Emory was waiting.

"Are you in a more cooperative mood?"

"I'll cooperate when you cops decide to treat my man and me like the good citizens we are. The reward isn't coming out of your pocket, so you have no right to withhold it. My man and I are simply claiming it like any other citizen would do."

"But we believe your man helped Jewel escape. He could be a felon, and you might be too for trying to collect ill-gotten money. We want to know your role in all of this."

'I'm going to tell the news media what you're doing."

"Does your man really know where Jewel is?"

"Except for Jewel himself, my man is probably the only person in the world who knows where he is. Tell me the truth. You got me out of the cell because a superior officer doesn't like your way of handling this issue. That officer deals with the public more than you do and is feeling the heat. Am I right?"

Hanlon's face turned red.

Detective Emory said. "We'll work with you."

The three men put together a plea bargain agreement.

"I'll take this agreement to my man. If he likes it, I'll give you a call."

***

Erick rented a room in the Downtown Value Inn, where he met with Conrad.

Erick handed Conrad the proposed agreement with the River Town Police Department. Conrad read it. Erick played the cassette recording to give Conrad the full picture.

Conrad said, "The cops recognize that I have information that will enable them to arrest Jewel. Make them confirm that Sherman Franklin will pay up. Once you get that assurance, go ahead and close the deal."

***

Erick met again with Hanlon and Emory at The Mad Greek. They made two cassette recordings.

Emory said, "I met with Sherman Franklin. He's enthused about the prospect of learning Jewel's whereabouts. He wanted us to sign the agreement, and so we did. Here's your copy."

Erick read the agreement and noted the Police Chief's signature.

"I'm glad you guys saw the light. Jewel is in Sydney, Australia. His alias is James Chase."

"Do you have an address or a phone number?" Emory asked.

"No, but Bryce attends operas and eats at Greek restaurants. Knowing his alias should help you find him."

"Do you know anything further?" Hanlon asked.

"The police in Sydney can check to see if he bought a house, even if he bought it under a name other than James Chase."

Emory said, "The U.S. and Australia have an extradition treaty, so I'll ask the police there to arrest Jewel."

***

Hanlon and Emory were at the River Town Police Department. Emory received a phone call.

"This is the U.S. Attorney's Office in Salem. Please hold for Carlos Ramirez, U.S. Attorney."

Ramirez came on the line.

"First, I assure you that this is a secure phone line. No one can tap this line."

Emory said, "Sir, I don't know if you will remember, but we met a few years ago. We were panelists at a law enforcement conference."

"I was getting around to mentioning that fact to you."

"I'm happy to work with you," Emory said.

"I'm investigating possible violations of federal laws at the Green State Medical College in River Town. The apparent violations involve both state and federal laws. We need to agree on how we can work together. We need to establish which of us is the lead agency and what our respective responsibilities are."

"I want to include Lieutenant John Hanlon in this conversation. He'll join us in a second."

Hanlon came on the line.

Ramirez said, "Since I am the top federal cop in Oregon, my office will lead the investigation. The psychiatry department of the medical college conducts clinical drug trials, which are regulated by the FDA, a federal agency. Even the state laws that may have been broken were done so as a result of the psychiatry department running drug trials, a federal responsibility. And whistleblowers brought the case to me under the federal whistleblower protection act.

"As the US Attorney, I should lead the investigation. I propose that your police department play a meaningful role, as it appears that grand larceny is involved. I have notified the campus police, and they're willing to participate on an investigatory team. Have I overlooked any agency that should be included?"

Detective Emory said, "What about the FDA? It's in charge of federal clinical drug laws."

"I want you to swear to secrecy before I answer that question."

Both Emory and Hanlon swore to keep confidential what Ramirez was about to say.

Ramirez said, "We will work with the FDA, but the whistleblowers who came to me reported what they believed to be violations of laws by the FDA. The agency may be too close to an offending drug company in this case. You must keep this secret and, at the same time, not jump to the conclusion that the FDA is guilty of wrongdoing. But the FDA will be a subject of this investigation."

Emory said, "As far as I can tell, you're covering all the bases. The River Town Police will be glad to follow your lead. I have news that I'm eager to share with you."

"Okay, we agree that I will lead the investigation. What's the news that you have?"

Emory said, "Bryce Jewel is in Sidney, Australia. That's according to an informant wanting to collect the reward that is offered for information leading to Bryce's conviction."

Emory described the meetings that he and Hanlon had held with Erick Taggart.

Emory said, "My jurisdiction is limited to River Town. I can't ask Australian officials to arrest Jewel. To satisfy the extradition treaty, you, a federal official, should take charge."

"I'll be glad to."

Chapter 47: I'm James Chase

It's a case of mistaken identity.
It's one big mistake.

--Steen Langstrupt
Metro, 2012

US Attorney Ramirez contacted the federal police in Sidney. He told them of Jewel's alleged crimes and that Jewel had assumed the identity of James Chase. The Sidney police began to observe operas and Greek restaurants as places that Bryce frequented.

Two cops picked up James Chase one night as he left the Sidney Opera House.

"We have reason to believe you're Bryce Jewel from the USA. "

"I'm James Chase."

"You look like Bryce Jewel. Look at this picture."

"So, there's a resemblance. I'm James Chase."

"Where did you go to primary school?"

"I don't have to answer that."

"You're right. But we have the responsibility of deciding if you're James Chase or Bryce Jewel. So, I ask again, where did you attend primary school?"

"I attended here in Sydney."

"What was the name of your first-grade teacher?"

“I want to get a lawyer.”

“You can make your phone call, but until your lawyer shows up, you’ll wait in the correctional centre.”

Bryce phoned a lawyer who told him that other than delaying the extradition, there wasn't anything that he could do for him. Bryce waived his rights and was extradited to River Town.

***

Bryce was in the River Town lockup. A jailer took him to a room where Lieutenant Hanlon was waiting. Bryce refused cigarettes but accepted coffee.

Hanlon smirked.

“So, you’re no longer James Chase. You’re Bryce Jewel again.”

“I want my lawyer here before we talk.”

Hanlon said to a jailer, “Take Mr. Jewel back to lockup.”

The jailer returned Bryce to his cell, where he waited until Attorney William Davis arrived. After Davis and Bryce talked, they met with Hanlon.

“So, you weren’t so smart after all. You were caught because one of your test subjects took his life. His brother offered a reward for information leading to your conviction.”

“The government can’t be thinking of charging me with being connected to the test subject’s death. I wasn’t even in the country when he died.”

"The decedent may have been too sick to be a test subject, and you were the one who enrolled him. Also, when you took off, you had to know that a Pharm Tech's ability to run safe drug trials would decline. I think the U.S. Attorney can show that you're complicit."

Chapter 48: Another Plea Bargain

> Plea bargaining accounts for almost 98 percent of federal corrections and 96 percent of state corrections in the United States
>
> --Lucian E. Dervan
> *Fourteen Principles and a Path Forward for Plea Bargaining Reform*
> American Bar Association, 2024

Davis said, "We want to make a plea bargain with you. Bryce is willing to testify against a silent partner who protected Pharm Tech from being exposed. Without Bryce's help, you might not convict that person."

"Tell me who the person is and I'll tell you if a plea bargain is possible."

"First, we make a deal, then Bryce will tell you what he knows."

"I could let your client sit in jail until he decides to cooperate."

"If we were a Banana Republic, you could do that. But we have a law requiring a speedy trial. I won't sit around while you play games. You need to decide if you want Pharm Tech's silent partner."

"Here's the standard form that my department uses for a plea agreement. Show me what you want on it."

Davis and Jewel marked up the form and gave it back to Hanlon. The lieutenant made changes. Davis and Jewel asked for things in return. The two sides reached an agreement.

Hanlon said, "You're not done. Detective Emory and the Chief of Police have to approve our deal."

"If they go hard against Bryce, the deal is off."

Hanlon left with the tentative agreement. He returned an hour later.

"Both men approved the deal."

Hanlon turned on a cassette recorder. Davis did the same.

"Raise your right hand. Do you swear to tell the truth, the whole truth, and nothing but the truth?"

"I do."

"We are at the Police Department in River Town, Oregon. I am Lieutenant John Hanlon. At 1:32 p.m. on May 22, 1996, we begin the deposition of PhD pharmacologist Bryce Jewel. He's co-owner, along with Dr. Richard Martinson, of Pharm Tech, which is a clinic that did drug trials for the Green State Medical College.

"Dr. Jewel, I'm asking you to identify the silent partner who protected the clinical drug-testing scheme that operated for about eight years, at the Green State Medical College."

Bryce looked at Attorney Davis, who nodded.

"The silent partner is Arnold Picket, the Chief Financial Officer of the medical college."

"What makes you believe he's the silent partner of interest to this deposition?"

"Psychiatrist Martinson often said so when he managed the clinic. Martinson met many times with Picket. Martinson and I put off decisions until Picket could review our proposals. Picket was a decision-maker. "

"How did Picket protect Pharm Tech?"

"He provided alternative explanations when anyone questioned the propriety of our trials. He would persuade the concerned person that Martinson was properly handling drug company proceeds."

"Did Picket have to defend Pharm Tech many times?"

"I don't know. He always dealt with Martinson, never with me. Picket wanted to limit how much I knew."

"Would you say that Picket took a share of the profit without having to do very much?"

"That's what Martinson said."

***

Arnold Picket was in his office at Green State Medical College. He received an intercom call from his secretary.

"Law enforcement officers are here to see you."

"Send them in."

Tommy Wong and John Hanlon came in.

Hanlon said, “You’re under arrest for stealing from the Green State Medical College. Stand and place your hands behind your back.”

Hanlon handcuffed Picket, read him his Miranda rights, and took him to the River Town jail.

Picket pleaded not guilty. The judge released Picket on his own recognizance. They knew each other from having taken courses together years earlier at the University of Oregon law school.

Chapter 49: Queen Berengaria

Though he be mad, there be method to it.

--William Shakespeare
Hamlet Prince of Denmark
Written sometime between 1599 and 1601

Ellen and Richard were in a conference room in Château Guillard.

Richard said, "Let me tell you about the woman who married King Richard in 1191. Her name was Berengaria."

"How long were they married?"

"For eight years, until 1199 when Richard died. He was hit by a crossbow bolt, and gangrene set in. I want to call you Berengaria."

"Tell me about her."

"Part of the time, Richard and Berengaria had a bad relationship. A Catholic priest ordered Richard to repair it. But there was love, too. Berengaria went with Richard during the first part of the Third Crusade. Later, when he was in prison, she did her best to raise ransom to get him released."

"I'll change my name if you'll stop restricting my use of the phone."

"I christen you Queen Berengaria."

***

Richard and Berengaria invited a Channel 2 TV crew to film their lives as royalty. They were waiting for the TV crew to arrive.

Berengaria said, “Are you sure you want to do this? We’re going to share our lives with strangers.”

“We’ll be careful about what they see.”

“Why let them see anything?”

“I’ll use the film to put us in the best light with the public. We want folks to love Lionheart and Queen Berengaria.”

“I still don’t understand.”

“I’m following the example of the pharmaceutical industry. It does what it wants. It persuades the public and regulators that treating hypothetical disorders as if they exist in nature is the right thing to do. The industry also sees the prescribing of vast amounts of psychotropic drugs. The industry does this by using propaganda. I plan to do the same kind of thing to avoid being held accountable for the way I ran Pharm Tech.”

Berengaria said, “The pharmaceutical industry has thousands of healthcare workers and many scientific journals to lie for them. How can you expect to do what the drug companies and psychiatrists do?”

“We’ll take the public back in time with King Richard and Queen Berengaria. Strong public support can do wonders.”

***

Two days later, in the evening, the couple watched the news special that Channel 2 had produced.

A commentator said, "Richard Martinson believes himself to be Richard the First of twelfth-century England. Martinson insisted that his girlfriend, Ellen Browning, change her name to Berengaria, who for eight years was Queen of England.

"Lieutenant John Hanlon of the River Town Police Department proposed cutting off the electricity, water, and sewer services. But Martinson has installed a generator, wells, and a septic system. Also, the activist group, "Live on Lionheart," said they will demonstrate against any effort to drive the king and queen from their castle."

The TV screen cut to crowds marching in the streets.

The commentator said, "Activists paraded in River Town, Portland, Eugene, and other cities the last two nights. Activists warned the Governor that he would not be re-elected if he allows tough measures to be taken against this royal couple. Richard intends to hole up in the castle for years."

***

A law enforcement task force met at the River Town Police Department. The force consisted of officers from the US Attorney's office, the FDA, the River Town Police Department, and the medical college's campus police.

U.S. Attorney Ramirez said, "The news media and the public are laughing at us. They say we're helpless. They don't understand how deadly clinical drug trials can be. They're fascinated by Martinson's claim that he's Richard the Lionheart reincarnated. They've fallen in love with the castle and the entry lions. They're living vicariously through King Richard and Queen Berengaria."

Hanlon said, "The TV special mocked Richard and Berengaria, but it mocked law enforcement more. It criticized me for suggesting that services like electricity and water be cut off to force Martinson to surrender. I don't like being mocked."

"How many of you watched the special?" Ramirez said.

All five officers raised a hand.

"How many felt the program insulted us?"

Five hands went up.

Hanlon said, "I'll offer a plan for taking this madman into custody. We can rule out any attack on the ground because there is a moat and a surrounding wall. And Richard has said he wants to pour boiling water on us. Also, his staff are armed with bows, crossbows, and swords. Of course, we'd win, but the public would blame us if there were casualties.

"But we could attack by using a helicopter and distractions. We could set off dynamite outside the moat, then another bomb in another location. We'd spread their attention so thin that they wouldn't know what to do. We'd set down a big helicopter filled with men. We'd overpower them."

Captain Tommy Wong said, "Richard is wed to twelfth-century weapons. So, if we use modern military weapons, the public will hate us. Don't underestimate how much people admire the Lionheart."

Hanlon said, "But we can't let Martinson live in the castle indefinitely."

Tommy Wong said, “I disagree. We must avoid bloodshed. River Town, the college, and the VA hospital are a community. We live together. If we harm the king, queen, or any of their servants, the public will make us pay. Martinson is no longer testing drugs, so the public doesn’t see the urgency that you do.”

***

Richard called a meeting with Berengaria and the servants. They were in the banquet hall.

“Cops have kept us pinned down for a week. They call me morning and evening, trying to get me to give up. My attorney is also negotiating with them. The cops have no idea how happy I am here.

“They don’t understand what I learned from running clinical drug trials. The pharmaceutical industry and psychiatry violate public health laws every day. They kill people when running their trials. What I learned is that you can break the law in plain sight of authorities like the FDA. Drug companies have proved you can lie your way out of any problem.

“I stayed put in River Town when the cops took after Pharm Tech. I used propaganda as the pharmaceutical industry does. I got the public to love King Richard and his queen just as the drug industry persuaded the public to love Prozac and Zyprexa. The drug companies seemed to be telling me to stay in place and rely on the court of public opinion.

“Look at Bryce. He fled the country but got caught and was brought back to stand trial. Look at Arnold, who stayed in place but didn’t get the public on his side. Like Bryce, he’s facing trial.

"I stayed in place but made it so the cops can't break in on me. I created an image that the public loves. But let's discuss the practical aspects of us living in harmony."

The chef, Milton Banister, said, "You bought geese to serve as guards, but they're also laying eggs. They symbolize what you're saying: 'be vigilant but go about your daily lives.'"

"It doesn't seem fair to take their eggs when they're earning their keep as sentries. Let the geese incubate their eggs. I want a bigger flock."

Rafael Mendoza, chief of farm operations, said, "With the warmer weather, the garden is thriving."

Milton said, "When you hired me, you said you wanted a chef willing to live away from society. But I hope we don't have to stay cooped up for long."

"My lawyer and I will work out a deal with the cops. For now, let's concentrate on making our lives here as pleasant as possible."

Milton said. "I agreed to this when you hired me, but the reality of the cops monitoring our every move is unnerving."

"I warned you."

"I'm not backing out. But after a couple of months, we will need to review our situation."

"Fair enough."

Chapter 50: Richard learns he is not Lionheart's descendant

Sometimes people don't want to hear the truth because they don't want their illusions destroyed.

--Friedrich Nietzsche
German philosopher
1844-1900

Richard lived a different life now that he had left Pharm Tech. Before, he spent his time bringing in money. Now he promoted Lionheart's story to increase the public's support of his stand against the law. He would tell of being a direct descendant of the ancient king. He set up a library within the castle on the history of early England. He intended to research the king's lineage and discover his own family tree.

He asked himself if his ancestry stemmed from a son or daughter of Richard the First.

Richard opened a history book and read that the Lionheart had no legitimate children. He learned that King Richard had a long-term mistress named Elezebet, and that she may have borne him as many as fifteen children, of which twelve lived to adulthood. The king acknowledged that one of them, Philip of Cognac, was his son.

"How can this be?" Richard asked Berengaria, "Dad told me that I am a direct descendant of the king. Now I learn that I am not related to King Richard unless my ancestor was illegitimate. I have no way to substantiate a claim that I am related to Lionheart."

"Your dad was trying to make you feel better," Berengaria said.

"But when I became an adult, why didn't he correct my misunderstanding?"

"Maybe it was a story passed down through generations of your family. For centuries, that's how people learned almost everything. Maybe the misinformation was passed on by Ethan's father or grandfather. Maybe sometime long ago, one of your ancestors stated this untruth, and no one ever thought to check it out. For example, you didn't check it out until today."

"How could I have been so stupid? Why didn't I do the tiny bit of research necessary to learn that the king had no legitimate children? Why did I accept Dad's word for my ancestry?"

"Probably because you wanted to believe that you are related to King Richard."

"Did Ethan lie to me to get me to stop daydreaming about King Arthur? Was that why Ethan misled me?"

"There's another possibility, of course, that you're an illegitimate heir. But according to your history books, none of Richard's illegitimate children bore his name."

"Don't mention that possibility to me. I can't go around bragging that I'm an illegitimate descendant of Richard the First."

"Well, while we're censoring topics, I say don't question your father's motives. He wasn't trained in literature or science, so he wasn't inclined to check for the truth of what a family member may have told him. He wanted the best for you."

"Maybe I should blame myself. Why didn't I check this out long ago? I work with the personal histories of patients all the time.

This is just another person's history--mine. It took only an hour to find out the truth."

"Does it really matter? So, you're not King Richard's great, great, great-grandson. What's the difference?"

"It calls into question whether I'm the Lionheart reincarnated."

"No, it doesn't. It reflects only on which version of family history you believe. To be reincarnated, it's not necessary for there to have been a traceable genetic heritage. What is necessary is for the reincarnated person to lead a virtuous life if the cycle of living and rebirth is ever to end."

"Is it too late for me to live an ethical life?"

"Not according to the religions that believe in reincarnation."

"What do I have to do?"

"Give up your quest for great wealth. Live a simple life. Help others avoid suffering. You know what to do. It's the doing that's the hard part."

Chapter 51: Richard regrets wrongdoing

> Confess you were wrong yesterday, and it will show you are wise today.
>
> Proverb

Berengaria was picking flowers in the courtyard. Penny Mordant, a servant, approached her.

"Pardon me if I don't call you Berengaria. I knew you as Ellen, and I don't buy this Queen Berengaria stuff."

"I don't either. Is there something I can do for you?"

"We servants want out. We'll board a helicopter if the cops can set one down in the courtyard."

"Is there something you want me to do?"

"You can phone the U.S. Attorney and speak for all of us."

"I'll do it. Richard loves living in the castle, but I don't. I miss my family. Is there anything more that I should know?"

"We servants respect Richard even though we want to get away from him."

"I feel the same. He suffered a brain injury when he was a boy. Maybe he wouldn't be so radical if he had not fallen and hit his head."

***

That evening, Berengaria spoke with Richard.

"Financial profit is not the only way to measure the success of a drug trial. To worship profit lowers the likelihood of scientific achievement."

"Do you think you're in a better position than me to judge the value of the Pharm Tech clinical trials?"

"You wrote papers that promoted the outcome of every trial. They can't all be as successful as you said. Let's consider why the cops are after you. And why they arrested Bryce and Arnold. Did you try to shift all the blame for dishonest drug trials onto them? It would be brave of you to admit you didn't publish the trials that showed that your favorite drugs performed poorly."

Finally, the argument ended with no clear winner. Richard had greater knowledge, but Berengaria had nothing to hide.

Richard's conscience began to bother him as he thought about what Berengaria had said.

He felt guilty about how he had built his fortune, how he had run Pharm Tech, and how he had been unfair to test subjects. He felt guilty about how he had dealt with Bryce and Arnold. He had robbed his employer, the medical college. The VA hospital had been kind to him, and he had stolen from that institution too.

As he reviewed his behavior, he thought of his sister. He felt the pain of Laura's suicide. He recalled how kind and gentle she had been. He wished he were more like her.

He had outdone other contract researchers, and that would have been good had he not cheated to do it. His recruiting of beautiful women to lure men into serving as test subjects now seemed wrong, whereas before it had seemed brilliant.

He was sending money to his mom and dad, and he felt good about it. He would have difficulty sending money had he not built a fortune. He was not about to give away his fortune, but he could be kind and decent for the rest of his life. That was the route he would take.

His thoughts returned to Laura. She protected nonhuman animals who had done nothing to harm humans, but whom humans killed with little regard. In trying to be more like Laura, Richard decided to turn Château Guillard and the wild acreage into a preserve for animals. He would provide a place where animals could live out their natural lives. He vowed he would stop eating any food derived from animals.

Richard was undergoing a change of character. One reason was the lectures that Berengaria had given him. Another reason was that he was no longer trying to build wealth. He was no longer obsessed with that goal. He had nothing left to prove to anyone, including himself. If only Laura were alive to guide him.

That night Richard slept fitfully. Laura visited him in his sleep. Was it a dream? The experience seemed so real.

Laura's image said, "Trillions of animals inhabit the earth. Humans regard themselves as more important than all non-human animals. Now that the public is focused on you, you can use your status to spread the word that humans and other animals are equal.

"By treating other animals with kindness, humans will automatically treat each other better."

Laura's image faded away.

Chapter 53: Richard stays where he belongs

> You must make a decision that you are going to move on. It won't happen automatically.
>
> —Joel Osteen
> American televangelist

Ellen received a phone call from her mother, Joyce.

"What will happen if the police raid the castle?"

"They won't raid as long as Richard is negotiating with them."

"How do you know that?"

"According to Richard, that's what the police told him."

"I wouldn't take his word. He's the one who got you guys in this fix. Ellen, what's next?"

"Don't worry, Mom."

It was early afternoon. Richard assembled his servants for crossbow practice. The target was a plywood human figure in front of a backstop. The group practiced for two hours.

***

Ellen phoned US Attorney Carlos Ramirez.

"This is Ellen Browning. Richard is out for a walk."

"Can I be of help?"

"The servants and I want to leave."

"What do you have in mind?"

Ellen described her plan. Carlos offered changes. They agreed on what to do.

***

Carlos assembled his team.

"Ellen Browning phoned me. She and her servants want to escape. They're willing to board a helicopter."

Tommy Wong said, "Helping them escape would be much easier than taking the castle by force."

Carlos said, "If we carry out this new plan, we can win the public over to our side. People will see that we're doing what the queen wants. We can save her and her staff and leave Richard behind. We can try him in absentia. If he is convicted, he could be sentenced to house arrest. That wouldn't be a perfect outcome, but we would avoid bloodshed."

"We'll maintain harmony in the community," Tommy said. "Let's do it."

***

Two nights later, Richard made his rounds of the castle, courtyard, and grounds. A servant was on guard atop the wall. Richard went back into the castle.

Berengaria served Richard a glass of wine. He drank it, then a second. He refused a third glass, kissed his queen, and went to bed.

Richard kept a key inside his bedroom to prevent being locked in. But on the other side of the door, Ellen pushed a chair up under the knob. Richard slept.

A helicopter set down in the courtyard. Squawking of the geese mixed with the roar of the helicopter. Richard woke up.

Lieutenant Hanlon jumped out of the helicopter so he could help Ellen and her staff board it.

Richard couldn't open the bedroom door. He backed up and charged, hitting it with his shoulder. The door flew open. The chair skittered across the floor.

Richard grabbed a crossbow and arrows. He went to a window. The queen and her servants were running toward the helicopter.

A gander flew at Hanlon, veering away at the last second. As the bird passed over the officer's head, he released a stream of feces that hit Hanlon in the face. A warm goo stung his eyes and made him choke. He stumbled and fell, landing on his back.

Geese surrounded him, beating their wings, hissing, squawking, and pecking. Hanlon turned over onto his hands and knees, got up, ran to the helicopter, and climbed in ahead of the escapees.

Ellen was within crossbow range. Richard decided who he would be for the rest of his life. He threw the crossbow and arrows to the floor.

Tommy Wong grabbed the runaways one at a time and pulled them into the helicopter. It lifted off. The squawking of the geese grew louder as the roar of the helicopter faded.

END

www.ingramcontent.com/pod-product-compliance
Lightning Source LLC
LaVergne TN
LVHW050910210926
848482LV00012B/390